D0162663

# Transition Metal Carbonyl Cluster Chemistry

## ADVANCED CHEMISTRY TEXTS

A series edited by DAVID PHILLIPS, *Imperial College, London, UK*, PAUL O'BRIEN, *University of Manchester, UK* and STAN ROBERTS, *University of Liverpool, UK*

**Volume 1**
Chemical Aspects of Photodynamic Therapy
*Raymond Bonnett*

**Volume 2**
Transition Metal Carbonyl Cluster Chemistry
*Paul J. Dyson and J. Scott McIndoe*

# Transition Metal Carbonyl Cluster Chemistry

Paul J. Dyson

*University of York, UK*

and

J. Scott McIndoe

*University of Cambridge, UK*

**Gordon and Breach Science Publishers**

Australia • Canada • France • Germany • India • Japan
Luxembourg • Malaysia • The Netherlands • Russia
Singapore • Switzerland

Amsteldijk 166
1st Floor
1079 LH Amsterdam
The Netherlands

**British Library Cataloguing in Publication Data**

ISBN: 90-5699-289-9
ISSN: 1027-3654

# Contents

# Foreword

Following the landmark discoveries of nickel tetracarbonyl and iron pentacarbonyl late in the 19th century, there was a period of development of the transition metal carbonyls which has led to a large and diverse chemistry virtually unprecedented in inorganic chemistry. This chemistry has led not only to a substantial knowledge of the reactivity of bonded carbon monoxide but also to numerous applications in catalysis and organic reactivity. Carbonyls have also played a significant and fundamental role in the development of transition metal cluster compounds and their applications.

This book is the first devoted solely to this enormously important area. It describes the whole spectrum of carbonyl chemistry from synthesis and characterisation to application in catalysis and synthesis. I am certain that this book will be essential reading for those either working in or contemplating working in this exciting area. It offers a firm and comprehensive basis for the subject and may serve as a text at both undergraduate and graduate level.

Brian F.G. Johnson FRS
Professor of Inorganic Chemistry and Master of Fitzwilliam College
Department of Chemistry
University of Cambridge, UK

# Preface

This book is part of a series of Advanced Chemistry Texts. It is intended to accompany undergraduate courses on transition metal carbonyl cluster chemistry. The subject is often taught on its own or as part of an advanced organometallic chemistry module. The book is aimed at final year BSc or MChem/MSc students although it will also be useful to students commencing research into cluster chemistry. The book is divided into two approximately equal parts; Chapters 1–5 deal with bonding, structure, spectroscopic properties and characterisation of clusters and Chapters 6–10 are concerned with synthesis, reactivity, reaction mechanisms and the use of clusters in organic synthesis and catalysis. Throughout the book we attempt to explain the unique properties of clusters that distinguish them from mononuclear carbonyl and other organometallic complexes. The text does assume some prior knowledge of organometallic chemistry.

We would like to take this opportunity to thank all those who have helped us with this book. We are particularly grateful to Professor Brian Johnson who has taken an interest in our project and has given us his full support, encouragement and advice. Brian Nicholson, Caroline Martin and David Brown have also made a significant contribution reading large portions of the book and helping to show us ways to present complicated ideas and information in a clear fashion. In addition, Catherine Judkins read parts of the book and made many useful comments.

# CHAPTER 1

# INTRODUCTION

This chapter commences by defining what actually constitutes a cluster in chemistry with respect to this book. It then proceeds to illustrate the different types of clusters encountered across the periodic table and places transition metal carbonyl clusters within this broader framework. Subsequently, attention is focused exclusively on transition metal carbonyl clusters. Their physical position between single metal complexes and the bulk metallic state is assessed, as is their chemistry, which is placed in context within the general area of organometallic chemistry. Lastly, the multi-functional roles that clusters play in catalysis is mentioned and cited as the main reason why cluster chemistry has witnessed such intensive research in the last few decades.

## 1.1   DEFINITION AND SCOPE

The clusters that are the subject of this book, i.e. transition metal carbonyl clusters, can be defined according to the following:

> A cluster is a molecule containing three or more metal atoms connected by direct metal-metal bonds.

However, there are many other types of compounds often referred to as clusters (see Figure 1.1). For example, there are many types of single element main group clusters such as $P_4$ (white phosphorus), $S_8$ and fullerenes (e.g $C_{60}$) as well as zintl ions, e.g. $Tl_9^{9-}$ and $Tl_{13}^{11-}$, that are formed by the late transition and post transition metals. Clusters that are composed of only skeletal atoms, i.e. with no other ligands or groups attached, like the main group compounds mentioned above, are comparatively rare. Most cluster skeletons are surrounded, at least to some extent, by a shell of ligands. This book deals with transition metal clusters that are primarily decorated with carbonyl ligands. However, other types of clusters are stabilised by different ligands including borane (hydroboron) clusters that from an historical perspective laid the foundations of cluster science, and $Fe_4S_4$ based clusters ubiquitous in biological electron transfer processes. Other examples of much larger main group-transition metal clusters include compounds such as $[Cd_{10}S_4Br_4(SC_6H_4Me-1,4)_{12}]^{4-}$, $Cu_{72}Se_{36}(PPh_3)_{20}$, $Cu_{146}Se_{73}(PPh_3)_{30}$ and $Ag_{30}Te_9(TePh)_{12}(PEt_3)_{12}$.

The compounds described above are defined as clusters as they have three-dimensional shapes and there are direct element-element bonds between the skeletal atoms. There are many more examples of cage compounds where three-dimensional

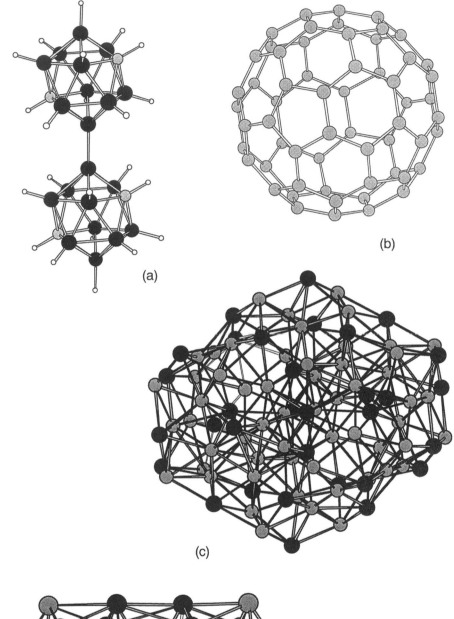

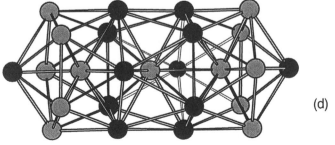

FIGURE 1.1 THE STRUCTURES OF (a) $\{C_2B_2H_{11}\}_2$, (b) $C_{60}$, (c) $Cu_{58}Te_{32}(PPh_3)_{16}$ AND (d) $[Au_{12}Ag_{13}(PPh_3)_{10}Br_8]^+$.

structures are formed without direct element-element bonds between the same element, and are therefore beyond the scope of this introduction.

## 1.2 TRANSITION METAL CARBONYL CLUSTERS

A common feature of the early transition metal clusters described above is that they interact with π-donor ligands, and hence, these clusters are termed π-donor clusters or high oxidation state clusters as the metals are in a high formal oxidation state. The transition metals in Groups 7–10 form clusters with π-acceptor ligands, most notably the carbonyl ligand. The metals are in low oxidation states and hence these clusters are either termed π-acceptor clusters or low oxidation state clusters. The π-acceptor ligands help to produce the most favourable condition for metal-metal bond formation of the Group 8–10 elements by inducing the greatest overlap between the atomic orbitals of the metals. In addition, overlap between atomic orbitals also increases as the principal quantum number increases. Therefore, as a transition metal sub-group is descended, the metal-metal bonds increase in strength.

With the π-acceptor clusters the metal-ligand and metal-metal combinations must also be kinetically inert towards bond dissociation otherwise fragmentation or colloid formation can take preference over clustering. The balance between metal-ligand and metal-metal bond strengths has allowed clusters with nuclearities ranging from three to around fifty to be isolated in the laboratory. This places clusters firmly in the region between mononuclear complexes and colloids and Figure 1.2 illustrates the progression in particle size from a single metal complex to the bulk metallic state.

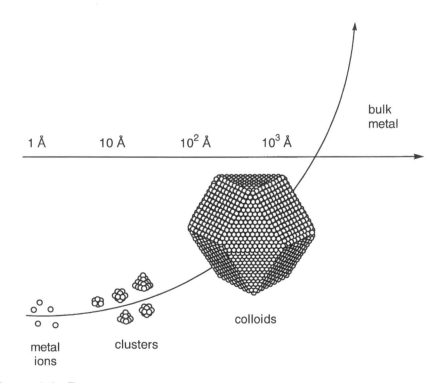

FIGURE 1.2  THE RELATIVE SIZE OF CLUSTERS COMPARED TO OTHER METAL STATES.

### 1.2.1 When Does a Cluster Become Metal-like?

As the size of a cluster increases it starts to behave less like an organometallic complex and more like a metal particle. There is no clear size at which a cluster becomes metal-like and certain clusters exhibit different types of metal-like properties under different conditions.

From a structural viewpoint, large clusters are closely related to bulk metals. Many large clusters have structures that can be considered as segments of close-packed metallic structures such as face-centred cubic packing, hexagonal close packing and body-centred cubic packing. For example, $[Rh_{15}(CO)_{27}]^{3-}$ has a structure that is a composite of body-centred cubic packing and face-centred cubic packing. These structural aspects are described further in Chapter 3. In addition, the interstitial sites within a bulk metal lattice can accommodate main group atoms and the same is true of clusters. Another feature of clusters that is comparable to bulk metals is their ability to form alloys. Chapter 9 is devoted to mixed-metal (heteronuclear) clusters and some of the largest clusters isolated so far are actually composed of mixed-metal systems.

High nuclearity clusters tend to be dark in colour compared to smaller clusters. The increasing intensity in colour with increasing nuclearity indicates that the separation between the highest bonding and lowest antibonding skeletal molecular orbital decreases with increasing cluster size. In fact, high nuclearity clusters have many closely spaced energy levels. There is a parallel between the d–d bandwidth of a metal particle (the band gap) and the separation of the highest occupied and lowest unoccupied skeletal molecular orbital (HOMO-LUMO gap) in a cluster and the two regimes are compared in Figure 1.3. While clusters do not have a conduction band like the bulk metal state, as the nuclearity of the cluster increases, the number of closely spaced energy levels also increases, and a wide range of different charge states can be achieved. For example, oxidation or reduction of some very large clusters can result in a cluster having ten different charge states.

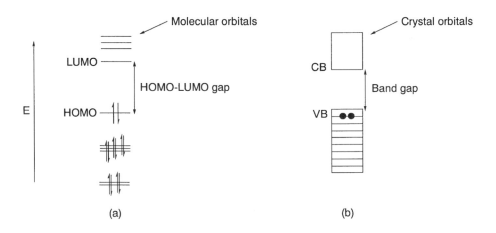

FIGURE 1.3 THE RELATIONSHIP BETWEEN (A) THE ENERGY LEVEL DIAGRAM OF A $M_6$ CLUSTER AND (B) BAND DIAGRAM OF A BULK METAL.

### 1.2.2 Growth in Transition Metal Carbonyl Cluster Chemistry

The chemistry of the π-acceptor clusters can be viewed as a specialised sub-section of π-acceptor (organometallic) chemistry. Cluster chemistry was preceded by mononuclear organometallic transition metal chemistry. Organometallic compounds were already known to stabilise organic molecules that could not otherwise be isolated, allowed novel organic synthesis to take place and were also known to be catalytically active. As the first metal carbonyl clusters were isolated in the laboratory and their structures established it became clear that they offered a new dimension to organometallic chemistry. It was found that ligands were able to coordinate to more than one metal centre and that this had a profound effect on the structure, spectroscopic properties and subsequent reactivity of the ligand. For example, as a carbonyl ligand interacts with an increasing number of metal centres, the C–O bond weakens (shown by IR spectroscopy and X-ray crystallography — see Chapter 4) to a point where it may undergo certain chemical transformations that are not possible when it bonds to just one metal centre. In addition, certain molecules can be stabilised by clusters that cannot be stabilised by mononuclear complexes, and both of the above factors also led to the proposal that clusters might exhibit novel catalytic functions. A scheme illustrating the roles played by clusters in catalysis is depicted in Figure 1.4.

Since clusters are regarded as a bridge between mononuclear complexes and colloids, but with similar solubilities to mononuclear complexes, they can be viewed as microscopic fragments of a bulk metal surface that can be studied using the same solution techniques as other homogeneous catalysts. It is in the area of catalysis that clusters have found most applications and Chapter 10 of this book is devoted to the subject of cluster catalysis.

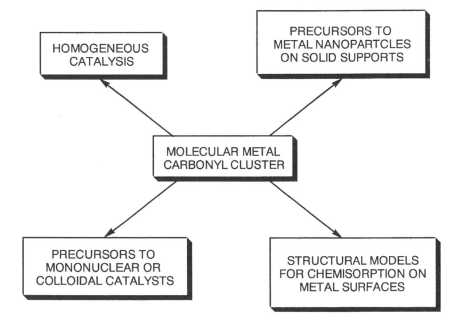

FIGURE 1.4  CLUSTERS IN CATALYSIS.

# CHAPTER 2

# ELECTRON COUNTING AND METAL-METAL BONDING

This chapter describes the methods used to rationalise and understand the structures of transition metal carbonyl clusters. There is not one single method that works for all compounds, but all are based on electron counting procedures. All the important methods are described, and it will become apparent where the different procedures should be applied. It will also be shown how certain clusters can be viewed in different ways, and in such cases, it is often possible to use more than one electron counting scheme to rationalise their structure.

## 2.1 THE LIGANDS

Transition metal carbonyl clusters often contain ligands other than CO and in order to count the electrons in a complex the number of electrons that a particular ligand donates must be known. It is a relatively straightforward task to work out the number of electrons donated by a given ligand using a simple valence bond approach. The number of electrons donated by the more commonly encountered ligands are listed in Table 2.1.

TABLE 2.1 NUMBER OF ELECTRONS DONATED BY LIGANDS IN THEIR MOST COMMON BONDING MODES.

| No. of electrons | Terminal | Bridging $\mu$- | Face-capping $\mu_3$- or $\mu_4$- | Interstitial |
|---|---|---|---|---|
| 1 | H, F, Cl, Br, I, $CR_3$, CN | H, $AuPR_3$ | H, $AuPR_3$ | H |
| 2 | CO, CNR, NCR, $NR_3$, $PR_3$, $SR_2$, $(\eta$-$C_2R_4)$ | CO, CNR, $CR_2$ | CO, CNR | |
| 3 | $(\eta^3$-$C_3R_3)$, NO | F, Cl, Br, I, $PR_2$, OR, SR, NO | CR, NO | B |
| 4 | $(\eta^4$-$C_4R_4)$ | $(\eta$-$C_2R_2)$ | O, S, PR | C |
| 5 | $(\eta^5$-$C_5R_5)$ | | F, Cl, Br, I | N, P |
| 6 | $(\eta^6$-$C_6R_6)$ | | $(\eta^2$:$\eta^2$:$\eta^2$-$C_6R_6)$ | O, S |

The Si, Ge and Sn equivalents of C based ligands donate the same number of electrons. The same rule applies to Se and Te variants of O and S donor ligands, and As and Sb variants of P donor ligands.

Certain ligands can bond to clusters in a variety of ways and only the most commonly encountered bonding modes are given in the Table. Ligand bonding modes are described in greater detail in Chapter 4.

---

### Bonding notation

The notation used to describe the way in which a ligand bonds to a cluster involves using the symbols $\eta^n$ and $\mu_m$ and sometimes $\sigma$. The symbol $\eta$ (hapto or eta) refers to the number of atoms (n) in the ligand which interact with the metal and by convention the superscript is not shown when all the atoms in the ligand are involved in bonding. The symbol $\mu$ (mu) refers to the number of metal atoms (m) to which the ligand bonds. This symbol is not used when the ligand bonds to only one metal atom and by convention the subscript is ignored if the ligand bonds to two metals ($\mu$ not $\mu_2$), but is included when the ligand bonds to three or more metal atoms ($\mu_3$, $\mu_4$ etc.). A bonding mode can be often be described in a number of different ways, all of which are correct, but different degrees of information are provided depending upon the precise nomenclature used. These ideas are illustrated below.

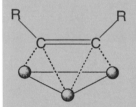

The bonding can be described as $\mu_3$-$\eta^2$:$\eta^1$:$\eta^1$ or $\mu_3$-$\eta^2$:$\sigma$:$\sigma$ and comprises one $\pi$-bond and two $\sigma$-bonds. Note that coordination in this manner results in elongation of the C-C bond and is therefore more like that of an alkene. The R-groups bend away from the trimetal face.

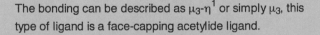

The bonding can be described as $\mu_3$-$\eta^1$ or simply $\mu_3$, this type of ligand is a face-capping acetylide ligand.

### 2.2 THE EIGHTEEN-ELECTRON RULE

The 18-electron rule states that 18 electrons are required from the metal and its associated ligands to attain a noble gas configuration. It is derived from the fact that transition metals have nine valence atomic orbitals [$5 \times nd$, $(n+1)s$ and $3 \times (n+1)p$], which can be used either for metal-ligand bonding or for the accommodation of

TABLE 2.2   THE NUMBER OF VALENCE ELECTRONS OF THE TRANSITION METALS IN THE ZERO OXIDATION STATE.

| Group | 3 | 4 | 5 | 6 | 7 | 8 | 9 | 10 | 11 | 12 |
|---|---|---|---|---|---|---|---|---|---|---|
|  | Sc | Ti | V | Cr | Mn | Fe | Co | Ni | Cu | Zn |
|  | Y | Zr | Nb | Mo | Tc | Ru | Rh | Pd | Ag | Cd |
|  | La | Hf | Ta | W | Re | Os | Ir | Pt | Au | Hg |
| s + d electrons | 3 | 4 | 5 | 6 | 7 | 8 | 9 | 10 | 11 | 12 |

non-bonding electrons. The formation of 18-electron carbonyl and organometallic complexes simply arises from the requirement to fill these nine orbitals, and hence gain the electronic configuration of the next highest noble gas (it is similar to the octet rule used to rationalise the structures of main group compounds). Transition metal complexes that do not contain $\pi$-acceptor ligands do not obey the rule and the reason for this different behaviour will become clear in due course. The simplest way to count the electrons about a single metal in a complex involves using the following set of rules:

i.    Always consider the metal atom (and all the ligands) to have an oxidation state of zero (the number of valence electrons for the transition metals in a zero oxidation state is given in Table 2.2).

ii.   Add together the valence electrons of the metal atoms and the electrons donated by all the ligands.

iii.  Account for any overall charge on the complex by adding (negative charge) or subtracting (positive charge) the appropriate number of electrons.

iv.   Consider a metal-metal single bond to provide one electron to each metal, a double bond to provide two electrons to each metal, etc. (as the number of metal atoms increases the metal-metal bonds cannot always be treated in this way — see below).

v.    Consider bridging ligands to provide an equal share of their electrons to each metal it interacts with.

The 18-electron rule is nearly always obeyed by the elements near the centre of the transition series. Their conformity arises because the energies of the nd atomic orbitals are close in energy to those of the (n+1)s and (n+1)p atomic orbitals and hence they all make a full contribution to the bonding. Further to the right of the transition series, in Groups 10 and 11, the nd orbital energies become significantly lower in energy than those of the (n+1)s and (n+1)p orbital, due to the increase in effective nuclear charge that has the greatest effect on the d orbitals, and these orbitals no longer make a full contribution towards bonding.

All the stable mononuclear homoleptic carbonyl complexes obey the 18-electron rule. Furthermore, it is easy to understand why $V(CO)_6$, which has only 17 valence electrons, is readily reduced to the 18 electron anion $[V(CO)_6]^-$.

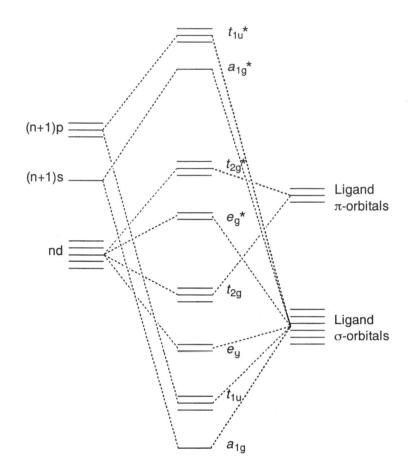

FIGURE 2.1 THE MOLECULAR ORBITAL ENERGY LEVEL DIAGRAM FOR $Cr(CO)_6$; NOTE THAT THE $a_{1g}$, $t_{1u}$, $e_g$ AND $t_{2g}$ MOLECULAR ORBITALS MUST BE FULL FOR A STABLE COMPLEX

From a molecular orbital point of view the 18-electron rule is best illustrated with reference to $Cr(CO)_6$ and the molecular orbital (MO) energy level diagram is shown in Figure 2.1. The 4s, 4p, $3d_z^2$ and $3d_{x^2-y^2}$ orbitals on the metal combine with the σ-donor orbitals of the six CO ligands, forming six MOs involved in M-L σ-bonding ($a_{1g}$, $t_{1u}$ and $e_g$ sets). The metal $t_{2g}$ orbitals, although σ non-bonding, are of the correct symmetry to participate in π-bonding, and back-donation of electron density occurs from the metal into the π-acceptor orbitals on CO (π*). Hence, there are nine molecular orbitals that have either σ- or π-bonding character (the $a_{1g}$, $t_{1u}$, $e_g$ and $t_{2g}$) and clearly 18 electrons are required to fill them.

The chromium atom is in a zero oxidation state and therefore provides six electrons while each CO ligand donates two electrons (a total of 12 electrons from the 6 ligands) giving a combined total of 18 electrons.

**Fe$_2$(CO)$_9$**

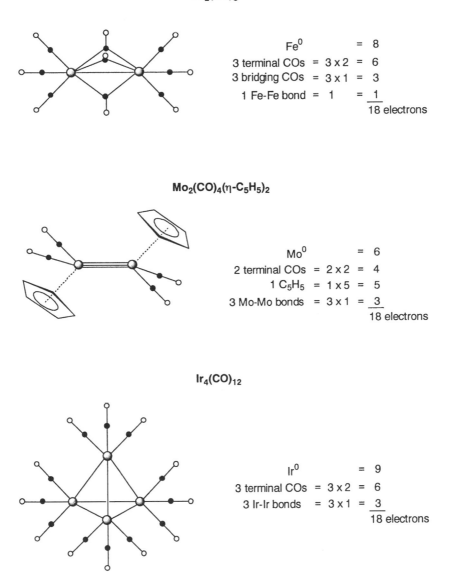

| | | |
|---|---|---|
| Fe$^0$ | | = 8 |
| 3 terminal COs | = 3 x 2 | = 6 |
| 3 bridging COs | = 3 x 1 | = 3 |
| 1 Fe-Fe bond | = 1 | = 1 |
| | | 18 electrons |

**Mo$_2$(CO)$_4$($\eta$-C$_5$H$_5$)$_2$**

| | | |
|---|---|---|
| Mo$^0$ | | = 6 |
| 2 terminal COs | = 2 x 2 | = 4 |
| 1 C$_5$H$_5$ | = 1 x 5 | = 5 |
| 3 Mo-Mo bonds | = 3 x 1 | = 3 |
| | | 18 electrons |

**Ir$_4$(CO)$_{12}$**

| | | |
|---|---|---|
| Ir$^0$ | | = 9 |
| 3 terminal COs | = 3 x 2 | = 6 |
| 3 Ir-Ir bonds | = 3 x 1 | = 3 |
| | | 18 electrons |

FIGURE 2.2 APPLICATION OF THE 18-ELECTRON RULE TO Fe$_2$(CO)$_9$, Mo$_2$(CO)$_4$($\eta$-C$_5$H$_5$)$_2$ AND Ir$_4$(CO)$_{12}$.

The application of the 18-electron rule is illustrated in Figure 2.2. For Fe$_2$(CO)$_9$, Mo$_2$(CO)$_4$($\eta$-C$_5$H$_5$)$_2$ and Ir$_4$(CO)$_{12}$, the metals in each compound are in equivalent coordination geometries and each metal obeys the 18-electron rule. In compounds containing metal-metal bonds the electron count about each metal atom must be calculated. In a dimer with two 18-electron centres, for example, the total electron count is 34 and not 36.

## 2.3 THE EFFECTIVE ATOMIC NUMBER (EAN) RULE

The EAN rule is a simple extension of the 18-electron rule. If the number of electrons associated with each metal in a cluster average 18, the cluster obeys the EAN rule. The rule is illustrated with reference to the osmium cluster $Os_6(CO)_{18}$ shown in Figure 2.3, which has a bicapped tetrahedral osmium core. This cluster contains six osmium atoms in three different coordination environments. $Os_a$ has three Os-Os bonds and three carbonyls, $Os_b$ has four Os-Os bonds and three carbonyls and $Os_c$ has five Os-Os bonds and three carbonyls.

**$Os_6(CO)_{18}$**

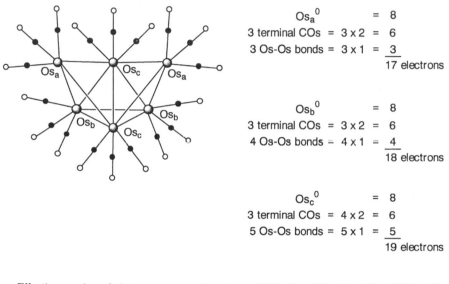

|                  |           |       |
| ---------------- | --------- | ----- |
| $Os_a{}^0$       |           | = 8   |
| 3 terminal COs   | = 3 x 2   | = 6   |
| 3 Os-Os bonds    | = 3 x 1   | = 3   |
|                  |           | 17 electrons |

|                  |           |       |
| ---------------- | --------- | ----- |
| $Os_b{}^0$       |           | = 8   |
| 3 terminal COs   | = 3 x 2   | = 6   |
| 4 Os-Os bonds    | = 4 x 1   | = 4   |
|                  |           | 18 electrons |

|                  |           |       |
| ---------------- | --------- | ----- |
| $Os_c{}^0$       |           | = 8   |
| 3 terminal COs   | = 4 x 2   | = 6   |
| 5 Os-Os bonds    | = 5 x 1   | = 5   |
|                  |           | 19 electrons |

Effective number of electrons per osmium atom = [(17 x 2) + (18 x 2) + (19 x 2)]/6 = 18 electrons

FIGURE 2.3   APPLICATION OF THE EAN RULE TO $Os_6(CO)_{18}$.

The pair of capping Os atoms, $Os_a$, are 'electron deficient' with only 17 electrons. One pair of the central Os atoms, $Os_b$, obey the '18-electron rule', while the other pair, $Os_c$, are 'electron rich' with 19 electrons. Clearly, in this example, the individual metal atoms do not obey the 18-electron rule, but if the total number of electrons are summed over the whole molecule and then divided by the number of metal atoms, then on average each osmium atom is associated with 18 electrons. Both the 18-electron rule and the EAN rule are based on the assumption that all the metal-metal (M-M) interactions consist of localised two centre/two electron (2c/2e) bonds and clusters which obey these rules may be termed 'electron-precise'. The bonding description in many larger clusters cannot be described in this way (see Section 2.4).

The EAN rule can be generalised by considering a cluster, $M_nL_x$, in which the metals and ligands do not have to be the same. The number of M-M 2c/2e bonds, $m$, is given by:

$$m = (18n - k)/2$$

where $k$ is the total electron count. Once again, the method can be illustrated with reference to $Os_6(CO)_{18}$:

$$Os_6(CO)_{18}: \quad 18n = 18 \times 6 = 108 \text{ electrons}$$
$$k = (6 \times 8) + (18 \times 2) = 84 \text{ electrons}$$
$$\underset{Os}{\phantom{k}} \qquad \underset{CO}{\phantom{k}}$$

Therefore, $m = (108 - 84) = 24$ electrons for M-M bonding, or alternatively 12 electron pairs, or 12 M–M 2c/2e bonds. The 12 metal-metal bonds predicted are in agreement with the 12 edges of the observed bicapped tetrahedron. The EAN rule is most frequently obeyed for clusters with a nuclearity of up to five metal atoms although some higher nuclearity clusters also obey the EAN rule. The trigonal prism represents another hexanuclear cluster that is electron precise and an example of a cluster with such a polyhedron is given below:

$$[Rh_6C(CO)_{15}]^{2-}: \quad 18n = 18 \times 6 = 108 \text{ electrons}$$
$$k = (6 \times 9) + (1 \times 4) + (15 \times 2) + 2 = 90 \text{ electrons}$$
$$\underset{Rh}{\phantom{k}} \qquad \underset{C}{\phantom{k}} \qquad \underset{CO}{\phantom{k}} \qquad \underset{charge}{\phantom{k}}$$

Therefore, $m = (108 - 90) = 18$ electrons for M-M bonding, or 9 electron pairs, or 9 M–M 2c/2e bonds. This conforms to a trigonal prism of rhodium atoms and is confirmed by the X-ray structure of the cluster shown in Figure 2.4. Clusters with very open metal frameworks also tend to obey the EAN rule.

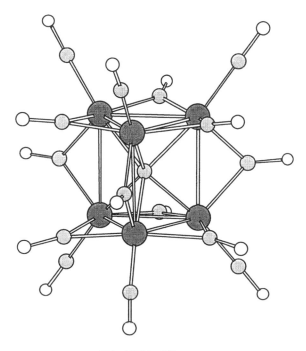

FIGURE 2.4  THE STRUCTURE OF $[Rh_6(CO)_{15}]^{2-}$.

The skeletal geometries of related 'electron-precise' clusters may be interconverted by the appropriate addition or removal of electron pairs. This type of process is believed to be important in certain catalytic reactions. The tetranuclear clusters $Os_4(CO)_{14}$, $Os_4(CO)_{15}$ and $Os_4(CO)_{16}$ are all electron precise and may undergo interconversion as shown in Scheme 2.1. They differ by the successive addition of the two-electron carbonyl ligand with the simultaneous cleavage of a metal-metal bond as each successive carbonyl is added. The geometry of the cluster core changes from tetrahedral for $Os_4(CO)_{14}$, to butterfly for $Os_4(CO)_{15}$, and finally to rectangular for $Os_4(CO)_{16}$.

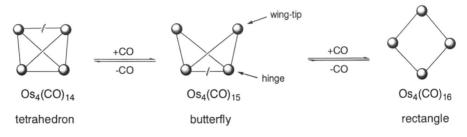

SCHEME 2.1 THE RELATIONSHIP BETWEEN THE 'ELECTRON PRECISE' CLUSTERS $Os_4(CO)_{14}$ (TETRAHEDRAL), $Os_4(CO)_{15}$ (BUTTERFLY) AND $Os_4(CO)_{16}$ (RECTANGULAR).

The butterfly cluster contains two different types of metals, denoted '*wing-tip*' and '*hinge*'. If the hinge-to-wingtip bond in $Os_4(CO)_{15}$ had been broken instead of the hinge-hinge bond then the resulting cluster formed would have a spiked-triangular metal core rather than a rectangular one. It is not possible to predict which geometry the cluster has using only the EAN rule and additional data is required to establish the actual geometry of the metal core. In some cases it may be possible to identify the metal atom most susceptible to nucleophilic attack, which may help to predict the structure of the cluster product. Similar interconversions of trinuclear and pentanuclear clusters can occur and for the larger cluster an even wider range of framework geometries are possible. It is worth noting that the addition of an electron pair to a cluster that is not electron precise does not always cause M-M edge cleavage but can result in a more fundamental structural reorganisation. An example of a polyhedral rearrangement involving a change from a bicapped tetrahedron to an octahedron takes place for $Os_6(CO)_{18}$ on reduction to $[Os_6(CO)_{18}]^{2-}$. This geometrical change can be accounted for using the polyhedral skeletal electron pair theory (see Section 2.6), which is not based on a 2c/2e M-M bonding model.

With few exceptions, in all the clusters of the iron and cobalt sub-groups the number of metal framework edges is always equal to or greater than the number of 2c/2e bonds predicted by the EAN rule. For clusters with six or more metal atoms the general tendency is for the clusters to have more edges than predicted. The characteristic electron counts of common polyhedra are listed in Table 2.3, which highlights the disparity between the observed number of bonds and the number

TABLE 2.3  THE CHARACTERISTIC ELECTRON COUNTS OF COMMON CLUSTER POLYHEDRA.

| Structure | $n$ | $E$ | $k$ | $m$ |
|---|---|---|---|---|
| Triangle | 3 | 3 | 48 | 3 |
| Linear/open triangle | 3 | 2 | 50 | 2 |
| Tetrahedron | 4 | 6 | 60 | 6 |
| Butterfly | 4 | 5 | 62 | 5 |
| Trigonal bipyramid | 5 | 9 | 72 | 9 |
| Square based pyramid | 5 | 8 | 74 | 8 |
| Edge bridged tetrahedron | 5 | 8 | 74 | 8 |
| Bridged butterfly | 5 | 7 | 76 | 7 |
| Bicapped tetrahedron | 6 | 12 | 84 | 12 |
| Octahedron | 6 | 12 | 86 | 11 |
| Capped square pyramid | 6 | 11 | 86 | 11 |
| Trigonal prism | 6 | 9 | 90 | 9 |
| Monocapped octahedron | 7 | 15 | 98 | 14 |
| Bicapped octahedron | 8 | 18 | 110 | 17 |
| Square antiprism | 8 | 16 | 114 | 15 |
| Tetracapped octahedron | 10 | 24 | 134 | 23 |

$n$ = number of metal atoms
$E$ = number of metal-metal bonds observed
$k$ = total electron count
$m$ = number of metal-metal 2c/2e bonds predicted
Note that as the nuclearity of the cluster increases beyond 6, the number of edges observed no longer correspond to that predicted using the EAN rule.

predicted by the EAN rule for clusters with more than six metal atoms. An alternative electron counting scheme is required to rationalise the bonding in most of the high nuclearity clusters.

## 2.4  WADE'S RULES

Localised bonding schemes are not always adequate to describe the bonding in transition metal clusters. A similar situation also applies to several classes of main group clusters including the boranes and carboranes. Borane (boron-hydrogen) clusters are often described as 'electron-deficient' because they form complexes that do not contain a sufficient number of electrons to provide 2c/2e bonds between all the linked atoms. From a historical perspective the discovery of boranes preceded transition metal clusters and a series of rules were developed to rationalise their structures. These rules were subsequently extended to account for the bonding observed in transition metal clusters, and can be summarised as follows:

An $n$ vertexed *closo* cluster will be held together by $n+1$ bonding electron pairs.
An $n$ vertexed *nido* cluster will be held together by $n+2$ bonding electron pairs.
An $n$ vertexed *arachno* cluster will be held together by $n+3$ bonding electron pairs.
An $n$ vertexed *hypho* cluster will be held together by $n+4$ bonding electron pairs.

The terms *closo*, *nido*, *arachno* and *hypho* correspond to a series of increasingly open deltahedra and it will become clear how to apply these rules in due course.

A *closo*-cluster is one in which all vertices are intact. A *nido*-cluster is derived from a *closo*-structure by removal of the vertex that has the highest connectivity, and so on. In other words, if a cluster atom is bonded to four others, then it will be removed in preference to one bonded to only three other cluster atoms.

Thus, a borane cluster with six atoms will have a *closo* or closed (*i.e.* octahedral) arrangement of boron atoms if it is held together by seven (*i.e. n*+1, where *n* is the number of boron atoms) bonding electron pairs. Consider the archetypal borane cluster, $[B_6H_6]^{2-}$. Each boron atom provides two electrons towards cluster bonding (the other valence electron of boron is used in forming a single bond with the hydrogen atom). When the six pairs of electrons from the boranes are combined with the two from the charge a total of seven electron pairs are present (i.e. 14 electrons) that can be used in cluster bonding. The predicted structure is an octahedron, which is consistent with the observed structure. The application of these rules to borane clusters is adequately described in most general inorganic chemistry textbooks and will not be dealt with any further.

### 2.5  THE ISOLOBAL PRINCIPLE

Wade's rules were initially developed to explain the bonding in the boranes and subsequently extended to rationalise the structures of the carboranes, metallocarboranes and ultimately transition metal clusters. Striking similarities were noted between the cluster bonding requirements of fragments such as BH, $CH^+$ and $M(CO)_3$ (M = Fe, Ru and Os). For example, two BH fragments in $[B_6H_6]^{2-}$ can be replaced by $CH^+$ fragments giving the neutral *closo*-cluster $B_4C_2H_6$. Two isomers of this compound exist, one in which the carbon atoms are *cis* and the other in which they are *trans* with respect to their location within the cluster skeleton. The similarity of the BH and $Ru(CO)_3$ fragments is illustrated by the clusters $[B_6H_6]^{2-}$ and $[Ru_6(CO)_{18}]^{2-}$. In the hexaruthenium cluster it is assumed that each ruthenium atom has three terminally coordinated carbonyl ligands.

The BH, $CH^+$ and $M(CO)_3$ fragments have the same bonding characteristics and are therefore said to be isolobal. The formal definition is:

> Two fragments are isolobal if the number, symmetry properties, approximate energy and shape of their frontier orbitals as well as the number of electrons occupying them are similar.

It is worth emphasising that although the frontier orbitals must be similar, they do not need to be identical. The isolobal principle provides an important bridge between inorganic and organic chemistry. The principle can be extended to include a wide range of fragments covering the whole area of organometallic chemistry and is very useful in determining the number of electrons that a particular $ML_n$ fragment contributes to cluster bonding. While the frontier orbitals of organic fragments are simply derived from VSEPR theory it is most convenient to describe the $ML_n$ fragments as 'pieces' of an octahedron as shown in Figure 2.5 and derive their frontier orbitals from the MO diagram of an octahedral complex as shown in Figure 2.6.

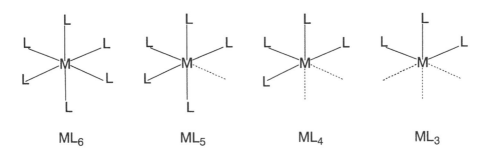

FIGURE 2.5  $ML_N$ FRAGMENTS AS 'PIECES' OF AN OCTAHEDRON.

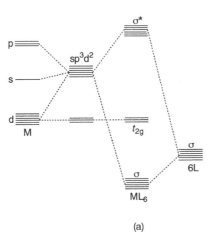

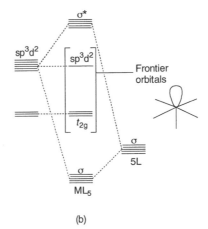

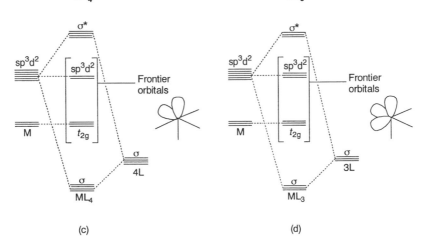

FIGURE 2.6  THE MOLECULAR ORBITAL ENERGY LEVEL DIAGRAMS FOR THE OCTAHEDRON $ML_6$ (a) AND THE FRAGMENTS $ML_5$ (b), $ML_4$ (c) AND $ML_3$ (d).

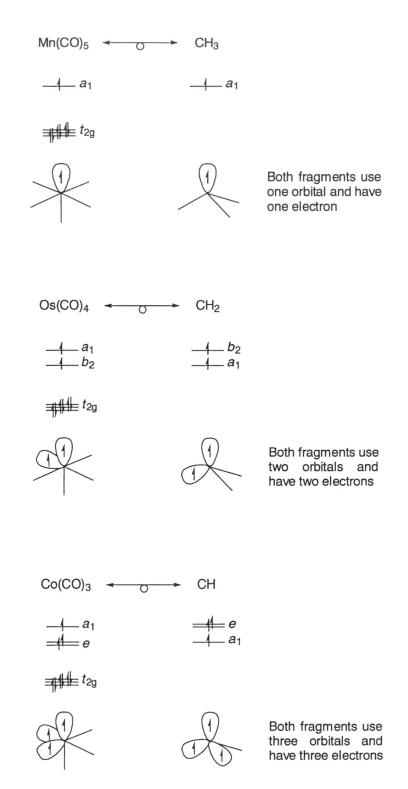

FIGURE 2.7 EXAMPLES OF THE ISOLOBAL RELATIONSHIPS BETWEEN TRANSITION METAL CARBONYL AND HYDROCARBON FRAGMENTS.

The orbitals used in cluster bonding can be derived from the MO diagram of an octahedral complex as shown in Figure 2.6(a). It can be seen from the diagram that six $sp^3d^2$ hybrid orbitals are generated by hybridisation of one s, three p and two d ($d_z^2$, $d_{x^2-y^2}$) metal atomic orbitals. These $sp^3d^2$ hybrids consist of single lobe orbitals, which point along the Cartesian axes and can interact with the appropriate ligand orbitals. There are also three unused d orbitals ($d_{xy}$, $d_{xz}$ and $d_{yz}$) and it is possible to construct the MO diagrams for the [ML$_6$] system. The important part of the diagram is the highlighted area that contains the frontier orbitals for the [ML$_5$] fragment, see Figure 2.6(b). Since only five ligand s-orbitals are available one of the six $sp^3d^2$ hybrid orbitals is left unused.

The frontier orbitals of the [ML$_4$] and [ML$_3$] fragments can be derived using the same principles applied to the [ML$_5$] fragment and they are also shown in Figure 2.6(c and d).

Some examples comparing metal carbonyl fragments to organic fragments with respect to the isolobal principle are illustrated in Figure 2.7. The symbol comprising a double-ended looped arrow is used to indicate which fragments are isolobal.

## 2.6 POLYHEDRAL SKELETAL ELECTRON PAIR THEORY (PSEPT)

The concept of isolobal fragments is useful when rationalising the structures of clusters such as Os$_4$(CO)$_{12}$($\mu_4$-C$_2$Me$_2$), which may be viewed as having a *pseudo-octahedral* skeleton formed by the four osmium atoms and two carbon atoms of the alkyne ligand (see Figure 2.8). It is easy to assign the number of electrons contributed by each part of the cluster and once the total number of electron pairs available for cluster bonding are known the structure can be predicted using Wade's rules.

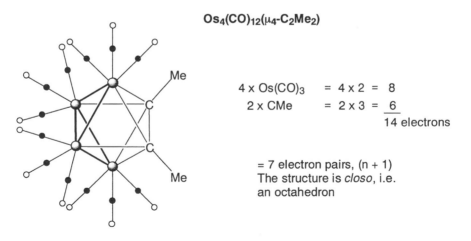

**Os$_4$(CO)$_{12}$($\mu_4$-C$_2$Me$_2$)**

| | | | |
|---|---|---|---|
| 4 x Os(CO)$_3$ | = 4 x 2 | = | 8 |
| 2 x CMe | = 2 x 3 | = | 6 |

14 electrons

= 7 electron pairs, (n + 1)
The structure is *closo*, i.e. an octahedron

FIGURE 2.8 THE APPLICATION OF THE ISOLOBAL PRINCIPAL TO THE RATIONALISATION OF THE ELECTRON COUNT FOR THE *PSEUDO*-OCTAHEDRAL Os$_4$C$_2$ CLUSTER Os$_4$(CO)$_{12}$($\mu_4$-C$_2$Me$_2$).

The structure of Os$_4$(CO)$_{12}$($\mu_4$-C$_2$Me$_2$) is based on a *closo* six-vertexed polyhedron. The electron count of this tetraosmium cluster can also be derived from the EAN rule, but only if the alkyne ligand is viewed as a six electron donor (four electrons donated from the π-orbitals and one electron from each of the s-bonds). Five metal-metal bonds are predicted which is consistent with the observed structure, i.e. a butterfly with a $\mu_4$-capping alkyne ligand.

Invoking the ideas developed for the isolobal principle, it is possible to extend Wade's rules from electron counting in borane clusters to transition metal clusters, as the above example shows. For transition metal clusters a general formula can be used in which the number of skeletal electron pairs, $S$, available for cluster bonding is given by:

$$S = (\text{Total number of valence electrons} - 12n) / 2$$

where $n$ is the number of skeletal metal atoms. The value 12 comes from the assumption that each skeletal metal atom uses six of its available atomic orbitals for metal-ligand bonding or for the occupation of non-bonding electrons leaving three available for framework bonding. This equation can be used to rationalise the structures of a large number of high nuclearity clusters and is exemplified in Figure 2.9 with $Fe_5C(CO)_{15}$ and $Ru_6C(CO)_{14}(\eta\text{-}C_6H_6)$.

### $Fe_5C(CO)_{15}$

Electron count: $(5 \times 8) + (1 \times 4) + (15 \times 2) = 74$
     Fe   C   CO

Number of skeletal electron pairs, $S = \dfrac{74 - (12 \times 5)}{2} = 7$

Five metal atoms are held together by 7 electron pairs (n + 2) so the polyhedron must be *nido*. This geometry is derived by taking a *closo*-cluster with one more metal atom (in this case an octahedron) and decapping to give a *nido*-cluster with one less metal, *i.e.* a square based pyramid.

### $Ru_6C(CO)_{14}(\eta\text{-}C_6H_6)$

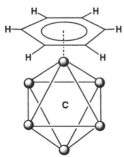

Electron count: $(6 \times 8) + (1 \times 4) + (14 \times 2) + (1 \times 6) = 86$
      Ru   C   CO   $C_6H_6$

Number of skeletal electron pairs, $S = \dfrac{86 - (12 \times 6)}{2} = 7$

Six metal atoms are held together by 7 electron pairs (n + 1) so the polyhedron must be *closo i.e.* an octahedron.

FIGURE 2.9 ANALYSIS OF THE ELECTRON COUNTS FOR $Fe_5C(CO)_{15}$ (*TOP*) AND $Ru_6C(CO)_{14}(\eta\text{-}C_6H_6)$ (*BOTTOM*) USING THE POLYHEDRAL SKELETAL ELECTRON PAIR THEORY.

## 2.7   THE CAPPING PRINCIPLE

As the nuclearity of transition metal clusters increases beyond six they do not tend to form larger *closo*-polyhedra. Instead, the frameworks are often based on smaller polyhedra in which one or more faces of the cluster are capped. Electron counting for such species is possible with the introduction of the capping principle, which states:

> A capped polyhedral cluster has the same number of electron pairs for framework bonding as the uncapped cluster.

The capping principle does not just apply to the capping of *closo*-polyhedra but also works for the capping of *nido-* and *arachno*-polyhedra, resulting in the following rules:

> A monocapped polyhedron with $n$ skeletal atoms has $n$ electron pairs available for framework bonding.
> A bicapped polyhedron has $n-1$ electron pairs available for framework bonding.
> A tricapped polyhedron has $n-2$ electron pairs available for framework bonding, *etc.*

This approach may be thought of as being equivalent to adding a capping group to a *closo*-cluster polyhedron without formally adding an electron pair. The frontier orbitals of the capping unit interact with orbitals that point out from the cluster face that is capped, and therefore these interactions do not disrupt the bonding in the cluster core. In order to apply the capping principle the number of skeletal electron pairs available for cluster bonding must be calculated using PSEPT. Application of the capping principle is illustrated in Figure 2.10 for the octanuclear cluster $Os_8(CO)_{23}$.

It should be noted that PSEPT, like the other electron counting schemes, does leave some ambiguities, and while it may be used as a means of predicting structures it will not differentiate between possible isomers or between say a *closo*-polyhedron and an equivalent capped *nido*-polyhedron. For example, there are three possible isomers of $Os_8(CO)_{23}$ in which different octahedral faces are capped. Similarly, if we consider the electron counts for $[Os_6(CO)_{18}]^{2-}$ and $H_2Os_6(CO)_{18}$ both have 7 electron pairs for framework bonding, and while $[Os_6(CO)_{18}]^{2-}$ is a regular

**$Os_8(CO)_{23}$**

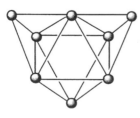

Electron count:  $(8 \times 8) + (23 \times 2) = 110$

$\qquad\qquad$ Os $\qquad$ CO

Number of skeletal electron pairs, $S = \dfrac{110 - (12 \times 8)}{2} = 7$

Eight metal atoms are held together by 7 electron pairs (n - 1) so the polyhedron must be a bicapped octahedron.

FIGURE 2.10   THE APPLICATION OF THE POLYHEDRAL SKELETAL ELECTRON PAIR THEORY TO THE *CIS*-BICAPPED OCTAHEDRAL CLUSTER $Os_8(CO)_{23}$.

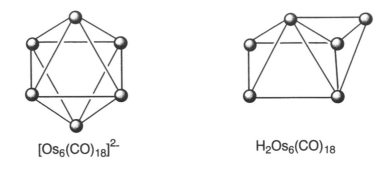

$$[Os_6(CO)_{18}]^{2-} \qquad H_2Os_6(CO)_{18}$$

FIGURE 2.11 THE STRUCTURES OF $[Os_6(CO)_{18}]^{2-}$ AND $H_2Os_6(CO)_{18}$.

octahedron, $H_2Os_6(CO)_{18}$ has a capped *nido*-octahedral framework as shown in Figure 2.11.

## 2.8 CONDENSED POLYHEDRA

A large number of high nuclearity carbonyl clusters have framework structures derived from the condensation of smaller polyhedra. They may be described as fused *closo*-polyhedra sharing a common vertex, edge or face. A simple electron counting approach for these clusters has been derived, which states:

> The total electron count in a condensed polyhedron is equal to the sum of the electron counts for the parent polyhedra minus the electron count characteristic of the atom, pair of atoms, or face of atoms common to both polyhedra.

The characteristic electron counts of some common cluster fragments are listed in Table 2.4. This method is very useful for predicting the total number of valence electrons which large clusters should have as long as it is possible to break the cluster down into smaller cluster units. Some applications of the condensation principle are shown in Figure 2.12.

TABLE 2.4 THE CHARACTERISTIC ELECTRON COUNTS OF UNITS COMMON TO BOTH POLYHEDRA.

| Shared unit | Subtract |
|---|---|
| Vertex | 18e |
| Edge | 34e |
| Triangular face | 48e |
| Triangular face | 50e[a] |
| Square face | 62e |
| Square face | 64e[b] |
| Butterfly face | 62e |

[a] When both parent polyhedra are deltahedra and have nuclearities of six or greater.
[b] When both parent polyhedra have M-M connectivities of three.

**Vertex sharing**

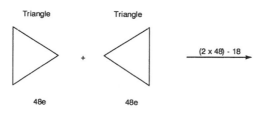
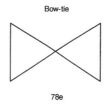

Triangle + Triangle → Bow-tie

$(2 \times 48) - 18$

48e    48e    78e

**Edge sharing**

Triangle + 3 [ Triangle ] → Raft

$(4 \times 48) - (3 \times 34)$

48e    48e    90e

**Face sharing**

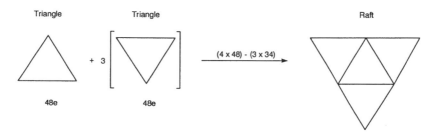

Tetrahedron + Tetrahedron → Trigonal bipyramid

$120 - 48$

60e    60e    72e

**Square face sharing**

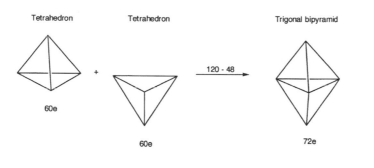

Trigonal prism + Trigonal prism → Face sharing trigonal prism

$180 - 64$

90e    90e    116e

FIGURE 2.12 EXAMPLE OF THE CONDENSATION PRINCIPLE INCLUDING VERTEX, EDGE AND TRIANGULAR/SQUARE FACE SHARING CLUSTERS.

# CHAPTER 3

# STRUCTURE

This chapter outlines the diverse range of structural types that transition metal carbonyl clusters adopt. Most attention is paid to the structures observed in the solid-state (which can differ from that in solution). While our attention is focused on clusters with three or more metal atoms the structures of mono and dinuclear complexes are briefly covered. The structure of the carbonyl shell is described for the smaller clusters but as the metal nuclearity increases most attention is directed towards the arrangement of the metal atoms.

### 3.1 MONONUCLEAR AND DINUCLEAR COMPLEXES

Stable mononuclear homoleptic carbonyl complexes are either volatile solids or liquids and the structures for the first row compounds are shown in Figure 3.1. Their structures are as expected for compounds with the coordination numbers they exhibit, *i.e.* tetrahedral, trigonal bipyramidal and octahedral for coordination numbers of four, five and six, respectively.

The carbonyl ligands in the tetrahedral and octahedral structures are all equivalent. However, the trigonal bipyramidal arrangement of carbonyl ligands in $Fe(CO)_5$ (see Figure 3.1b) gives rise to carbonyl ligands in two different environments; two axial and three equatorial. Using $^{13}C$ NMR spectroscopy it has been found that in solution all the carbonyl ligands in $Fe(CO)_5$ are equivalent at room temperature, as the $^{13}C$

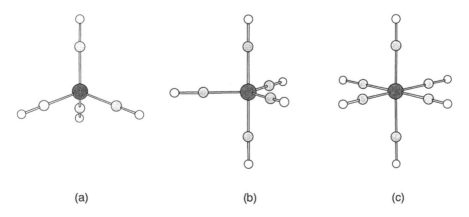

(a)                    (b)                    (c)

FIGURE 3.1   THE STRUCTURES OF THE HOMOLEPTIC CARBONYLS (a) $Ni(CO)_4$, (b) $Fe(CO)_5$ AND (c) $Cr(CO)_6$.

NMR spectrum shows only one resonance. Equivalence of all the carbonyl ligands on the NMR timescale is consistent with a Berry *pseudo*-rotation about the iron centre and the mechanism for this process is illustrated in Scheme 3.1. This type of ligand motion is known as stereochemical nonrigid behaviour, fluxionality or carbonyl scrambling and is extremely widespread in metal carbonyl compounds. Other examples of this phenomenon are given in Chapter 5.

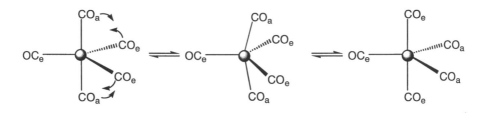

Fe(CO)$_5$ (CO$_a$ = axial, CO$_e$ = equatorial)

SCHEME 3.1   THE BERRY *PSEUDO*-ROTATION OF Fe(CO)$_5$.

The dinuclear homoleptic carbonyls are important from a historical perspective. The structural determination of Fe$_2$(CO)$_9$ in 1938 provided the first example of close metal atoms in a carbonyl compound (~2.5 Å) and later the structure of Mn$_2$(CO)$_{10}$ showed beyond doubt the presence of a metal-metal bond in a molecular compound as the bond was not supported by any bridging ligands. The Group 7 dimers M$_2$(CO)$_{10}$ (M = Mn, Tc and Re) adopt staggered conformations in the solid-state as shown in Figure 3.2a. In solution each M(CO)$_5$ unit can rotate about the metal-metal single bond. Fe$_2$(CO)$_9$ is the only stable dimer of the Group 8 sub-group and its structure is shown in Figure 3.2(b). Three terminal carbonyl ligands are bonded to each iron atom and the Fe–Fe bond is bridged by three carbonyls. The structure of Co$_2$(CO)$_8$ is less regular. One might anticipate four terminal carbonyl ligands bonded to each cobalt atom, which is indeed one of the structures observed at ambient temperatures in solution. However, in the solid-state there are three terminal carbonyl ligands bonded to each cobalt atom and two carbonyls bridge the Co-Co bond as shown in Figure 3.2(c).

### 3.2   TRINUCLEAR AND TETRANUCLEAR CLUSTERS

Much effort was devoted to establishing the structures of the trinuclear clusters M$_3$(CO)$_{12}$ (M = Fe, Ru and Os) and the mixed-metal species M$_2$M'(CO)$_{12}$ for which all possible combinations have been prepared and characterised. The species containing all three Group 8 metals, i.e. FeRuOs(CO)$_{12}$, has yet to be isolated. The trimetal cores of all these clusters are triangular, the arrangement of carbonyl ligands differs from cluster to cluster.

The structure of Fe$_3$(CO)$_{12}$ differs from that of Ru$_3$(CO)$_{12}$ and Os$_3$(CO)$_{12}$. In the last two species all the carbonyl ligands are terminal, four connected to each metal atom as shown in Figure 3.3. The solid-state structure of Fe$_3$(CO)$_{12}$

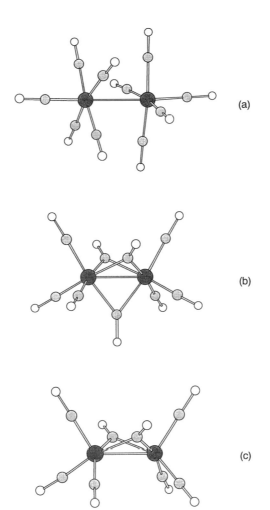

(a)

(b)

(c)

FIGURE 3.2 THE STRUCTURES OF (a) $Mn_2(CO)_{10}$, (b) $Fe_2(CO)_9$ AND (c) $Co_2(CO)_8$.

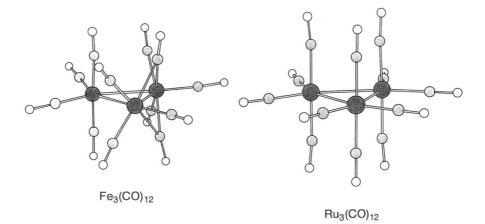

Fe₃(CO)₁₂

Ru₃(CO)₁₂

$Fe_3(CO)_{12}$

$Ru_3(CO)_{12}$

FIGURE 3.3 A COMPARISON OF THE STRUCTURES OF $Fe_3(CO)_{12}$ AND $Ru_3(CO)_{12}$; THE STRUCTURE OF $Os_3(CO)_{12}$ IS ANALOGOUS TO THAT OF THE RUTHENIUM CLUSTER.

contains two bridging carbonyl ligands along one Fe-Fe edge with the remaining carbonyls occupying terminal sites. It is not possible to predict where bridging ligands are likely to occur. In this case, the difference between the carbonyl distributions of the iron cluster and the ruthenium and osmium clusters is considered to be of steric origin as the shorter Fe-Fe bond is more easily bridged by a CO ligand than the longer Ru-Ru or Os-Os edges. However, there are many examples of ruthenium and osmium clusters that do contain edge bridging carbonyl ligands. It is worth noting that the bridging carbonyl ligands in $Fe_3(CO)_{12}$ do not bridge the Fe-Fe bond symmetrically although as the temperature at which the structure is determined is lowered, the bridging carbonyls become more symmetrical. In solution, the $^{13}C$ NMR spectrum of $Fe_3(CO)_{12}$ shows only one resonance at room temperature. Even if it has a structure like that of the ruthenium and osmium analogues two peaks should still be observed since there are two distinct carbonyl environments. The presence of only one peak in the NMR spectrum suggests that all the carbonyl ligands in $Fe_3(CO)_{12}$ are equivalent due to scrambling of the carbonyls. The mechanism by which the scrambling occurs is still not known with certainty despite many studies involving a wide range of different analytical and spectroscopic techniques, although one possible mechanism is described in Chapter 5.

The homoleptic tetranuclear clusters $M_4(CO)_{12}$ (M = Co, Rh and Ir) are all based on a tetrahedral metal core. In terms of their carbonyl ligand coverage the simplest of the three is the iridium cluster which has twelve terminal carbonyl ligands, three bonded to each iridium atom. The cobalt and rhodium clusters each have three bridging carbonyls around one face of the cluster. The remaining nine carbonyls are all terminal as shown in Figure 3.4.

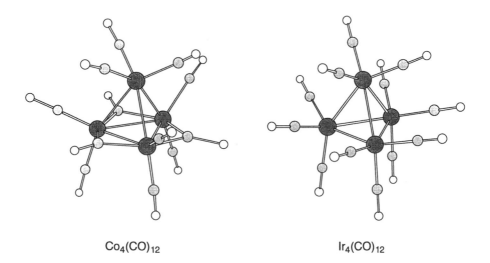

$Co_4(CO)_{12}$        $Ir_4(CO)_{12}$

FIGURE 3.4 A COMPARISON OF THE STRUCTURES OF $Co_4(CO)_{12}$ AND $Ir_4(CO)_{12}$; THE STRUCTURE OF $Rh_4(CO)_{12}$ IS ANALOGOUS TO THAT OF THE COBALT CLUSTER.

Clearly, as the number of metal atoms increases the number of possible structures also increases. With trinuclear clusters, apart from a triangular metal core, breaking one of the metal-metal bonds will result in a linear (or bent) arrangement of metal atoms. For a tetranuclear species several structural types may exist including the tetrahedron and butterfly, which are the most commonly observed shapes, as well as the rectangle and spiked triangle.

## 3.3    PENTANUCLEAR AND HEXANUCLEAR CLUSTERS

From a historical perspective, clusters with five or more metal atoms are often referred to as *high nuclearity* since it is from this point that the metal-metal bonds often deviate from two centre/two electron bonds and the 18-electron or effective atomic number rules tend to break down. However, today there are so many much larger clusters the term is used less frequently to describe clusters with five or six metal atoms and is more commonly used to describe clusters with nuclearities in excess of 20 metal atoms.

The two main geometries observed for pentanuclear clusters are the trigonal bipyramid and the square pyramid, these being *closo-* and *nido*-structures respectively, according to PSEPT (see Chapter 2). For hexanuclear clusters the most common geometries are the octahedron and the trigonal prism. The former represents a *closo*-structure and the latter, which is not a deltahedron (i.e. composed of all triangular faces), is electron precise and therefore obeys the EAN rule. These four common structural types are illustrated in Figure 3.5.

The structure of the octahedral cluster $Rh_6(CO)_{16}$ is shown in Figure 3.6. It represents one of the earliest high nuclearity clusters to be fully characterised by X-ray crystallography. The X-ray structure showed that four of the carbonyl ligands cap alternate faces of the octahedron. The face-bridging bonding mode for CO was unique at the time the structure was established. The remaining 12 carbonyls are all terminal, two bonding to each metal atom.

The hexaosmium dianion $[Os_6(CO)_{18}]^{2-}$ is a particularly interesting octahedral cluster since it undergoes a facile transformation upon oxidation to give the neutral cluster $Os_6(CO)_{18}$, which has a different metal atom geometry as shown in Scheme 3.2. The structure of $Os_6(CO)_{18}$ is best described as a capped trigonal bipyramid or a bicapped tetrahedron. The latter is preferred due to the aesthetic appeal of three tetrahedral units fused in a row.

Clusters with five and six metal atoms (or greater) can also contain interstitial main group atoms as the cavity formed by the metals is sufficiently large for them to be accommodated. Interstitial atoms are also found in tetranuclear butterfly clusters situated between the wing-tip metal atoms, however, the atom is exposed and often participates in reactions. The cavity in a trigonal bipyramid is too small to house an interstitial atom whereas the square pyramid, octahedron and trigonal prism all have cavities of sufficient size to accommodate a small main group atom. The radius of the cavity in an octahedral cluster is 0.414 times the metallic radius of the metal atoms on the surface of the cluster. Therefore only small main group atoms such as H, C and N may be incorporated into this site. Trigonal prismatic

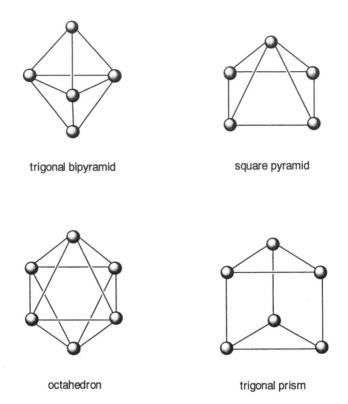

trigonal bipyramid　　　　　　　square pyramid

octahedron　　　　　　　trigonal prism

FIGURE 3.5　COMMON POLYHEDRA FOR PENTANUCLEAR AND HEXANUCLEAR CLUSTERS.

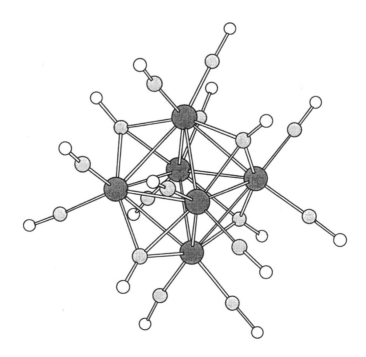

FIGURE 3.6　THE STRUCTURE OF $Rh_6(CO)_{16}$.

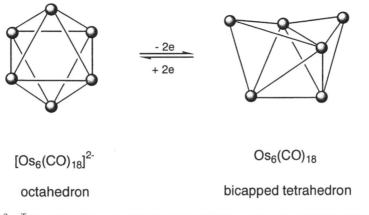

$$[Os_6(CO)_{18}]^{2-}$$

octahedron

$$Os_6(CO)_{18}$$

bicapped tetrahedron

SCHEME 3.2 THE OCTAHEDAL ↔ BICAPPED TETRAHEDRAL SKELETAL REARRANGEMENT.

clusters have a slightly larger cavity, 0.528 times the metallic radius, due to the presence of square faces and they may accommodate both first and second row main group interstitial atoms. Completely encapsulated transition metal atoms are also observed in much larger cavities created by higher nuclearity clusters and certain clusters contain more than one interstitial atom. Examples of these will be given throughout the book.

The interstitial atom can move slightly within the cavity of the cluster. An example where this takes place is in the three closely related clusters $Ru_6C(CO)_{17}$, $Ru_6C(CO)_{14}(\eta-C_6H_6)$ and $Ru_6C(CO)_{11}(\eta-C_6H_6)_2$ shown in Figure 3.7. These clusters have octahedral cores containing a carbide atom. In $Ru_6C(CO)_{17}$ there are 16 terminal carbonyl groups and one bridging CO and the interstitial carbon atom lies directly in the centre of the octahedral cavity. In $Ru_6C(CO)_{14}(\eta-C_6H_6)$ the CO ligand distribution is very similar to that of $Ru_6C(CO)_{17}$ except that on one ruthenium atom three carbonyls have been replaced by a benzene ligand. Since benzene is a less efficient π-acceptor than the three carbonyls it has replaced, the carbide atom moves closer to the ruthenium atom to which the benzene is coordinated. In the *bis*-benzene cluster $Ru_6C(CO)_{11}(\eta-C_6H_6)_2$ the $Ru_6$ octahedron is sandwiched between two benzene rings, which coordinate to opposite ruthenium atoms. The entire octahedron is compressed along the benzene-benzene axis further illustrating the compensatory effect of the carbide atom.

## 3.4 CAPPED POLYHEDRA

As transition metal clusters increase in size the metal skeletons do not tend to form larger and larger *closo*-polyhedra or *closo*-derived polyhedra (although some are known). Instead smaller polyhedra such as the tetrahedron or octahedron tend to be capped. Progressively capping more and more faces of a central cluster polyhedron eventually gives rise to some very large clusters. For example, capping a face of the octahedral cluster $[Os_6(CO)_{18}]^{2-}$ affords the heptaosmium cluster $[Os_7(CO)_{20}]^{2-}$ whose structure can be rationalised using the capping principle (see Chapter 2).

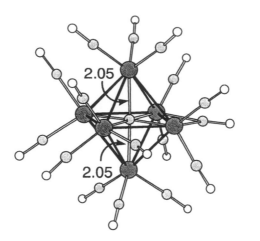

(a)

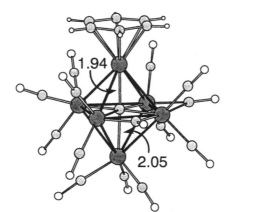

(b)

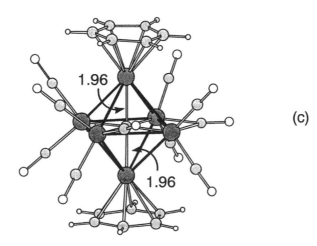

(c)

FIGURE 3.7 THE STRUCTURES OF (a) $Ru_6C(CO)_{17}$, (b) $Ru_6C(CO)_{14}(\eta\text{-}C_6H_6)$ AND (c) $Ru_6C(CO)_{11}(\eta\text{-}C_6H_6)_2$ SHOWING THE CONTRACTION OF THE Ru-CARBIDE BOND LENGTH WHEN A BENZENE REPLACES THE CARBONYL LIGANDS ON A Ru-ATOM.

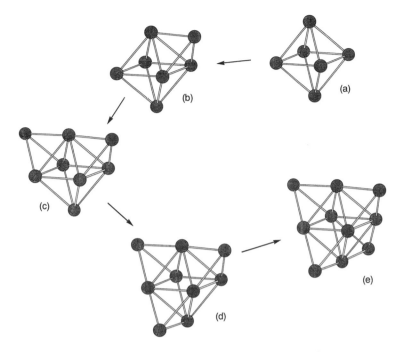

FIGURE 3.8 THE STRUCTURES OF (a) $[Os_6(CO)_{18}]^{2-}$, (b) $[Os_7(CO)_{20}]^{2-}$, (c) $[Os_8(CO)_{22}]^{2-}$, (d) $[Os_9(CO)_{24}]^{2-}$ AND (e) $[Os_{10}C(CO)_{24}]^{2-}$ SHOWING THE RELATIONSHIP OF THE METAL CORES BY SEQUENTIALLY CAPPING TRIANGULAR FACES OF $[Os_6(CO)_{18}]^{2-}$.

Capping a second face results in the octanuclear cluster $[Os_8(CO)_{22}]^{2-}$. It should be noted that these compounds have not actually been prepared in this stepwise fashion, but are isolated from a range of different reactions, they have been grouped together due to their obvious structural relationships. Continuing with this sequence, capping $[Os_8(CO)_{22}]^{2-}$ affords the nonaosmium cluster $[Os_9(CO)_{24}]^{2-}$, and if one further triangular face is capped then the decanuclear cluster $[Os_{10}C(CO)_{24}]^{2-}$ is obtained. In contrast to the other clusters in this series $[Os_{10}C(CO)_{24}]^{2-}$ contains an interstitial carbide atom in the central octahedral cavity. This sequence of clusters is shown in Figure 3.8.

The decaosmium cluster $[Os_{10}C(CO)_{24}]^{2-}$ can also be viewed as having a large pyramid-type structure and as such is part of the pyramidal (or tetrahedral) growth sequence illustrated in Figure 3.9. The first member of this sequence is the tetrahedral cluster $[Os_4(CO)_{13}]^{2-}$. The addition of six osmium atoms to the base of $[Os_4(CO)_{13}]^{2-}$ produces the metal core observed in decaosmium cluster $[Os_{10}C(CO)_{24}]^{2-}$. The next cluster in the sequence is formed by adding a layer of ten osmium atoms to the decaosmium cluster thereby affording the twenty metal atom cluster $[Os_{20}(CO)_{40}]^{2-}$. These three clusters follow a cubic close-packing growth sequence with the metal atoms stacking in layers according to the magic number sequence (1 : 3 : 6 : 10 *etc.*). The next layer in this series would contain 15 osmium atoms, giving rise to a cluster containing 35 metal atoms. There is some evidence to suggest that an

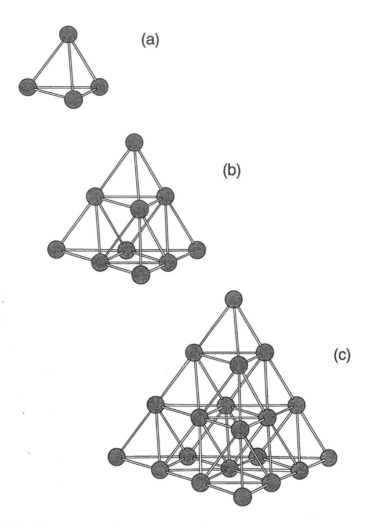

(a)

(b)

(c)

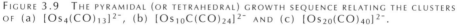

FIGURE 3.9 THE PYRAMIDAL (OR TETRAHEDRAL) GROWTH SEQUENCE RELATING THE CLUSTERS OF (a) $[Os_4(CO)_{13}]^{2-}$, (b) $[Os_{10}C(CO)_{24}]^{2-}$ AND (c) $[Os_{20}(CO)_{40}]^{2-}$.

osmium cluster containing this number of osmium atoms has been prepared although full characterisation has yet to be obtained. It is worth noting that while the clusters $[Os_4(CO)_{13}]^{2-}$, $[Os_{10}C(CO)_{24}]^{2-}$ and $[Os_{20}(CO)_{40}]^{2-}$ can be viewed as segments of a cubic close-packing lattice the actual structure of metallic osmium is not CCP. The comparison of metal clusters as fragments of metallic lattices will be expanded in Section 3.7.

It is also possible for metal fragments to cap non-triangular cluster faces such as rectangular and butterfly faces. Two examples of a clusters in which rectangular faces have been capped are $[Rh_9P(CO)_{21}]^{2-}$ and $[Rh_{10}S(CO)_{22}]^{2-}$. These clusters are derived from an octarhodium square antiprismatic structure, which contains two rectangular faces. In $[Rh_9P(CO)_{21}]^{2-}$, one rectangular face has been capped, and in

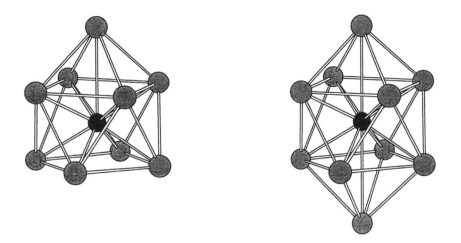

$[Rh_{10}S(CO)_{22}]^{2-}$, both of the rectangular faces are capped. The structures of $[Rh_9P(CO)_{21}]^{2-}$ and $[Rh_{10}S(CO)_{22}]^{2-}$ are illustrated in Figure 3.10.

The observation that clusters can grow by addition of metal fragments to pre-formed *closo-* and *nido-*clusters has been exploited in the laboratory for the synthesis of mixed-metal (*hetero*nuclear) clusters. Specific ligands may also be introduced into a cluster by attachment of the ligand to a metal fragment, which then condenses with a cluster, resulting in the simultaneous increase in the nuclearity of the cluster as well as the introduction of the new ligand (or ligands). Some examples of these types of reaction are given in Chapter 9.

### 3.5   COUPLED AND CONDENSED POLYHEDRA

In the previous section the sequential capping of metal fragments to a cluster polyhedron to form larger clusters was described. It is also possible for actual clusters to couple together to form larger polyhedra through coupling or condensation through a vertex, an edge or a face, and some examples are given below.

Certain clusters may couple with the formation of a metal-metal single bond. For example, the rhodium cluster $[Rh_{12}(CO)_{30}]^{2-}$ is composed of two octahedral $Rh_6$ units connected via a Rh-Rh single bond as shown in Figure 3.11. The bond connecting the two octahedra is bridged by two carbonyl ligands. The remaining carbonyl ligand distribution is not dissimilar from that of $Rh_6(CO)_{16}$ in that there are four face-capping carbonyl ligands on alternating faces of the cluster and two terminal carbonyls attached to each rhodium excluding the ones involved in the bridge. $[Rh_{12}(CO)_{30}]^{2-}$ is prepared by reducing $Rh_4(CO)_{12}$ and the mechanism of its formation probably involves the formation of $Rh_6$ octahedral clusters, which combine to form $[Rh_{12}(CO)_{30}]^{2-}$. In a more predictable synthesis, the octanuclear

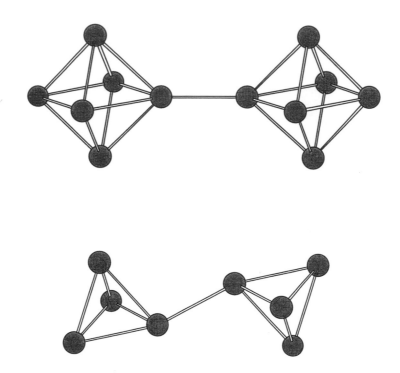

FIGURE 3.11: THE STRUCTURES OF $[Rh_{12}(CO)_{30}]^{2-}$ (*TOP*) AND $[Ir_8(CO)_{22}]^{2-}$ (*BOTTOM*) ILLUSTRATING COUPLING OF CLUSTERS THROUGH A VERTEX.

cluster $[Ir_8(CO)_{22}]^{2-}$ (see Figure 3.11) is prepared from the reduction of $Ir_4(CO)_{12}$, which brings about the coupling of the two $Ir_4$ units.

A series of nickel and platinum clusters of general formula $[M_3(CO)_6]_n^{2-}$ (M = Ni, n = 2 and 3; M = Pt, n = 2, 3, 4 and 5) can be viewed as triangular stacks in which the $M_3$ triangles couple through their faces. The structure of the platinum cluster where n = 5, *i.e.* $[Pt_{15}(CO)_{30}]^{2-}$ is shown in Figure 3.12. The triangles tend to stack on top of each other in an eclipsed conformation giving rise to a column of slightly twisted trigonal prisms. The metal-metal bonds within the triangular units are much shorter than those between the metal-metal bonds in the layers. Variable temperature $^{195}Pt$ NMR spectroscopy shows that the triangular units undergo rotation relative to each other so that there is a rapid interchange between trigonal prismatic and trigonal antiprismatic structures. The solid-state structures of the analogous nickel clusters differ from the platinum compounds. In $[Ni_6(CO)_{12}]^{2-}$ the structure is distorted such that the two triangular units adopt a trigonal antiprismatic structure as shown in Figure 3.12. In the triple-layered cluster $[Ni_9(CO)_{18}]^{2-}$ the central triangle unit forms a trigonal prismatic polyhedron with one of the $Ni_3$ units and a trigonal antiprismatic polyhedron with the other.

Certain clusters may also be viewed as being constructed from condensed polyhedral geometries in which two (or more) polyhedra are not linked by a metal-metal

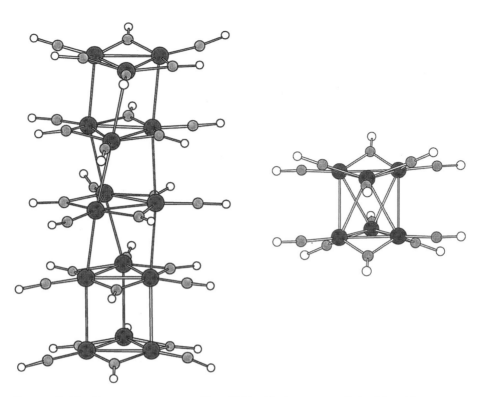

FIGURE 3.12 THE STRUCTURES OF $[Pt_{15}(CO)_{30}]^{2-}$ (*LEFT*) AND $[Ni_6(CO)_{12}]^{2-}$ (*RIGHT*).

bond (or bonds) but through shared atoms. For example, the heptaosmium cluster $[Os_7(CO)_{20}]^{2-}$ shown in Figure 3.8 above could be considered to be composed of an octahedron condensed with a tetrahedron to form the capped octahedral cluster observed. While this is an entirely reasonable way of viewing the structure of $[Os_7(CO)_{20}]^{2-}$, it is perhaps simpler to consider it as a capped octahedron. There is often more than one way to view the structure of a cluster and recognising the simplest description is helpful when deciding which electron counting technique to apply to rationalise the bonding (see Chapter 2). Condensed clusters (unlike coupled clusters) cannot physically be broken down into their simpler building blocks as the atoms that link the individual polyhedra are shared with each other. For example, the skeleton of the decanuclear cluster $[Ru_{10}C_2(CO)_{24}]^{2-}$ shown in Scheme 3.3 may be described as two edge-sharing octahedra. The metal core in $[Ru_{10}C_2(CO)_{24}]^{2-}$ undergoes a transformation when reacted for a prolonged period at elevated temperature with diphenylacetylene (PhC $\equiv$ CPh) to afford $[Ru_{10}C_2(CO)_{22}(\mu\text{-}C_2Ph_2)]^{2-}$, which has an additional bond formed between two apical ruthenium atoms on adjacent octahedra as shown in the Scheme; the alkyne ligand bridges this new edge. This new cluster can therefore be viewed as a tricondensed cluster involving two octahedra and one tetrahedron.

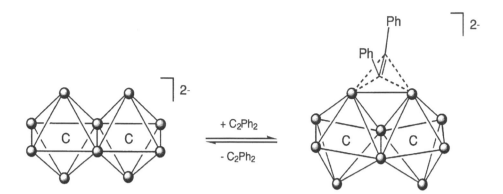

SCHEME 3.3 THE TRANSFORMATION OF $[Ru_{10}C_2(CO)_{24}]^{2-}$ TO $[Ru_{10}C_2(CO)_{22}(\mu\text{-}C_2Ph_2)]^{2-}$
ON REACTION WITH DIPHENYLACETYLENE.

An example of a cluster which is composed of two octahedral cluster units that
share a common trimetal face is $[Ir_9(CO)_{19}]^{3-}$ (see Figure 3.13).

The trigonal bipyramid could also be viewed as having a face-sharing tetrahedral
structure. However, the simplest description is usually the best one when describing
the geometry of a cluster.

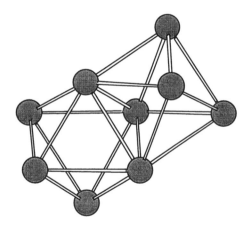

FIGURE 3.13 THE STRUCTURE OF $[Ir_9(CO)_{19}]^{3-}$, REPRESENTING AN EXAMPLE OF TWO
OCTAHEDRAL CLUSTERS CONDENSED THROUGH A TRIANGULAR FACE.

## 3.6 OPEN AND PLANAR CLUSTERS

A large number of relatively open deltahedral clusters are known. Some are approximately planar and these are often termed rafts. Open clusters tend to obey the EAN
rule even when they contain more than four metal atoms. Scheme 3.4 shows how
it is possible (at least theoretically) to progress from a *closo*-trigonal bipyramidal

structure to afford a raft and a so-called 'bow-tie' cluster. These transformations are brought about by the addition of two electron donor ligands, which causes the stepwise breaking of metal-metal edges. Perhaps the most notable examples of raft clusters are $Os_6(CO)_{20}(NCMe)$ and $Os_6(CO)_{19}(NCMe)_2$, which are isolated from the pyrolysis of $Os_3(CO)_{10}(NCMe)_2$. Since $Os_3(CO)_{10}(NCMe)_2$ is also an important precursor to high nuclearity *closo*-clusters, it is thought that these raft clusters may be intermediates on route to the larger, closed clusters.

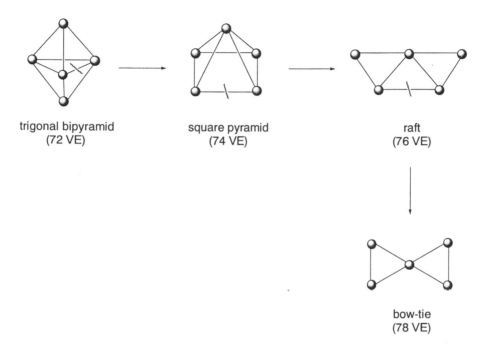

SCHEME 3.4 THE POSSIBLE TRANSITION FROM A *CLOSO*-TRIGONAL BIPYRAMIDAL STRUCTURE TO A RAFT AND BOW-TIE CLUSTER VIA SEQUENTIAL METAL-METAL BOND CLEAVAGE.

## 3.7 LARGER CLUSTERS

As clusters increase in size their properties become more metal-like and they begin to exhibit physical characteristics that are similar to those of small metal particles. Careful analysis of the metal atom structure of a large cluster often reveals that it is analogous to that of a metallic lattice and can therefore be described as a fragment of a bulk metal. In this respect it was mentioned in Section 3.4 that $[Os_4(CO)_{13}]^{2-}$, $[Os_{10}C(CO)_{24}]^{2-}$ and $[Os_{20}(CO)_{40}]^{2-}$ can be viewed as segments of a cubic close-packed lattice. Many large clusters are difficult to categorise into any of the structural types described so far. It is best to break down the cluster into simpler structural motifs and one common way in which this is performed is to analyse the cluster layer by layer. Consider, for example, the rhodium cluster $[Rh_{28}N_4(CO)_{41}H_x]^{4-}$ shown in Figure 3.14. This cluster may be viewed as being composed of three layers containing atoms in an A/B/C sequence. Four nitrogen atoms and an unknown number of hydrogen atoms occupy sites within (or over) the metal atom core. Due to the large size of $[Rh_{28}N_4(CO)_{41}H_x]^{4-}$ and the presence of so many heavy rhodium atoms, the comparatively small and light hydrogen atoms are not distinguished by X-ray diffraction and hence the precise number of hydrides remain uncertain as shown in the formula. This problem is often encountered in large clusters composed of second or third row transition metals and definitive structural characterisation requires a neutron diffraction study (see Chapter 5). There are many clusters that are similar in size to $[Rh_{28}N_4(CO)_{41}H_x]^{4-}$ or even larger. They are prepared by a range of techniques and the structure of the actual product is seldom predictable and characterisation relies heavily on X-ray crystallography. Some of the largest clusters are composed of two different transition metals and examples of these are given in Chapter 9.

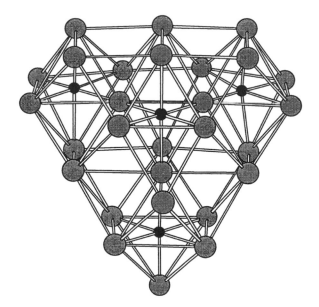

FIGURE 3.14 THE STRUCTURE OF $[Rh_{28}N_4(CO)_{41}H_x]^{4-}$; NOTE, FOR CLARITY NOT ALL Rh-Rh BONDS ARE INCLUDED.

### 3.8 CLUSTERS WITH METAL-METAL MULTIPLE BONDS

All of the clusters described in the preceding sections are coordinatively saturated whether they obey the EAN rule or the polyhedral skeletal electron pair model. Certain clusters do not appear to obey these rules and close inspection of their X-ray structures or careful analysis of their characteristic reactivity patterns, generally indicate metal-metal multiple bonds. One well documented example of a cluster with an apparent Os=Os double bond is $H_2Os_3(CO)_{10}$ shown in Figure 3.15. This cluster has a total electron count of 46, which is two fewer than predicted for a triangular

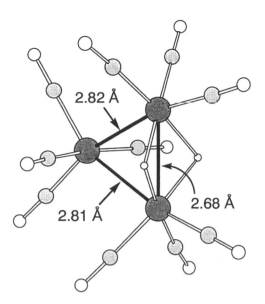

FIGURE 3.15 THE STRUCTURE OF $H_2Os_3(CO)_{10}$ SHOWING THE SHORT OSMIUM-OSMIUM BOND BRIDGED BY THE TWO HYDRIDE LIGANDS.

cluster. The X-ray structure reveals that the two hydride ligands bridge one edge of the cluster and that this edge is shorter than the others despite being bridged by the two hydride ligands, which usually cause lengthening of a metal-metal edge. The reactivity of this cluster is also typical of a compound containing an unsaturated bond in that it undergoes addition reactions instead of substitution reactions towards two electron donor ligands such as phosphines.

Reaction of the *bis*-acetonitrile cluster $Os_6(CO)_{16}(NCMe)_2$ with two equivalents of $Pt(\eta^4\text{-}C_8H_{12})_2$ affords the heteronuclear cluster $Os_6Pt_2(CO)_{16}(\eta^4\text{-}C_8H_{12})_2$. The cluster skeleton of $Os_6Pt_2(CO)_{16}(\eta^4\text{-}C_8H_{12})_2$ consists of two osmium tetrahedra fused along an edge with the two platinum atoms capping the faces of one of the tetrahedral units (Scheme 3.4). Using the PSEPT an electron count of 110 would be predicted for $Os_6Pt_2(CO)_{16}(\eta^4\text{-}C_8H_{12})_2$, but the cluster only has 108 CVE. Analysis of the X-ray structure shows that one of the Os-Os edges is abnormally short [2.65 Å] and constitutes a double bond, which would account for the cluster having two electrons fewer than predicted. Again, the unsaturated nature of $Os_6Pt_2(CO)_{16}(\eta^4\text{-}C_8H_{12})_2$ is demonstrated by its reactivity. In solution, $Os_6Pt_2(CO)_{16}(\eta^4\text{-}C_8H_{12})_2$ reacts immediately with CO at room temperature to form $Os_6Pt_2(CO)_{17}(\eta^4\text{-}C_8H_{12})_2$. This process may be reversed by heating $Os_6Pt_2(CO)_{17}(\eta^4\text{-}C_8H_{12})_2$ in octane for a few minutes regenerating $Os_6Pt_2(CO)_{16}(\eta^4\text{-}C_8H_{12})_2$ and involves a polyhedral change since the metal atom structure of the saturated cluster differs from that of the unsaturated cluster as shown in Scheme 3.5.

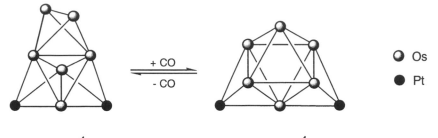

$Os_6Pt_2(CO)_{16}(\eta^4-C_8H_{12})_2$ $\qquad$ $Os_6Pt_2(CO)_{17}(\eta^4-C_8H_{12})_2$

SCHEME 3.5 THE TRANSFORMATION OF THE UNSATURATED CLUSTER $Os_6Pt_2(CO)_{16}(\eta^4-C_8H_{12})_2$ TO THE SATURATED CLUSTER $Os_6Pt_2(CO)_{17}(\eta^4-C_8H_{12})_2$ ON UPTAKE OF CO; NOTE THE PROCESS IS REVERSIBLE. NOTE THE $C_8H_{12}$ LIGANDS ARE OMITTED FOR CLARITY.

# CHAPTER 4

# LIGANDS

Low oxidation state transition metal clusters are surrounded and stabilised by a shell of ligands. The ligands that are encountered are generally much the same as those observed in mononuclear organometallic complexes. Carbon monoxide, phosphines, nitric oxide, hydrides, $\pi$-bonded ligands such as alkenes, alkynes, cyclopentadienyls and arenes are all common ligands in cluster chemistry. However, these ligands often display a much richer coordination chemistry when bound to clusters, as many bonding modes are available in which the ligands can interact with more than one metal centre.

## 4.1  CARBON MONOXIDE

Carbon monoxide is the most important ligand in transition metal carbonyl cluster chemistry. It is a versatile ligand, displaying a range of bonding modes, and has the important property of stabilising metals in low oxidation states and its small size means that a relatively large number can surround a metal cluster. The bonding between a metal and carbon monoxide nearly always involves coordination via the carbon atom, and is made up of two components:

i.      Donation of the lone pair of electrons on the carbon atom to an empty metal d orbital of $\sigma$ symmetry.
ii.     Back donation from filled metal d orbitals to the empty $\pi^*$ C–O antibonding orbitals.

These two bonding components are shown in Figure 4.1.

Carbon monoxide is a very poor $\sigma$ donor. Despite this, CO is a good ligand because the back-donation of electron density into the C–O $\pi^*$ orbitals increases

FIGURE 4.1    THE TWO COMPONENTS OF THE METAL–CARBONYL BOND.

the ability of the CO to donate electron density to the metal, which in turn increases back donation, and so on. The σ and π components work co-operatively, and this type of bonding is termed *synergic*. The donation of electron density into the C–O π* molecular orbital has two main effects on the M–C≡O linkage. First, the π bonding strengthens the M–C bond, resulting in partial double bond character. Direct measurement of the bond lengths show a distinct shortening of the M–$C_{CO}$ bond, with M–$C_{CO}$ distances typically 0.2–0.3 Å shorter than pure M–C σ bonds. Second, the C–O bond is weakened by the introduction of electron density into the antibonding π* orbitals. Structural studies show slightly longer C–O bond lengths, but the most sensitive method for detecting the C–O bond strength is IR spectroscopy. Free CO has a carbonyl stretching frequency, ν(CO), of 2143 cm$^{-1}$ and coordination to a metal reduces this value, often substantially. The extent to which the C–O bond is weakened depends on several factors and these will be covered in the following sections.

### 4.1.1 Bonding Modes of CO

Three CO bonding modes are commonly observed in clusters. Most compounds have mainly terminal CO ligands, but edge-bridging (μ-CO) and face-capping (μ₃-CO) modes are also frequently encountered. For the purposes of electron counting all three bonding types are two-electron donors.

The anionic cluster [Fe₃(CO)₁₁]²⁻ contains all three types of CO ligand (see Figure 4.2) and is fully represented by the formula [Fe₃(μ-CO)(μ₃-CO)(CO)₉]²⁻. For convenience, the different bonding modes are not generally distinguished in this way, especially when other ligands are present.

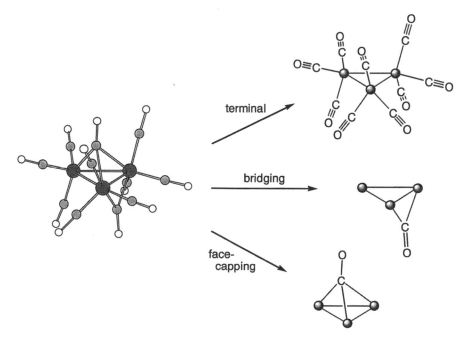

FIGURE 4.2    THE STRUCTURE OF [Fe₃(μ-CO)(μ₃-CO)(CO)₉]²⁻ SHOWING THREE COORDINATION MODES OF CO.

## Unusual bonding modes of CO to clusters

In addition to terminal, bridging ($\mu$), and face-capping modes ($\mu_3$), CO may bond to clusters in a variety of other ways. These bonding modes (illustrated below) are rare.

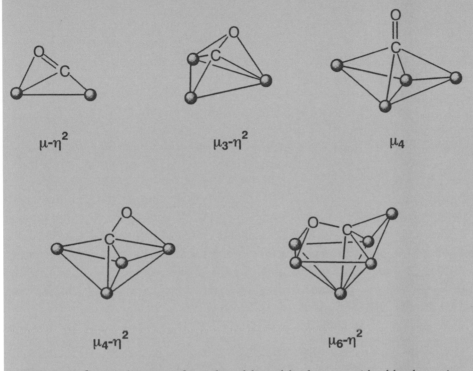

Side-on ($\eta^2$) coordination of a carbonyl ligand leads to considerable elongation and weakening of the CO bond. Further stretching of the C–O bond ultimately leads to cleavage with the formation of carbide and oxo species, an important reaction that is discussed in more detail in Chapter 7.

Bridging carbonyls are better $\pi$-acceptors than terminal CO ligands because there is more effective overlap between the d orbitals of two metals with the $\pi^*$ orbitals of the carbonyl. The increased $\pi^*$ back-bonding manifests itself clearly in the $\nu(CO)$ values, which are considerably lower than those of terminal ligands. This effect is even more pronounced for face-capping CO ligands. Typical CO stretching frequencies for the three modes are shown in Table 4.1.

TABLE 4.1

|  | free | terminal | bridging ($\mu$) | face-capping ($\mu_3$) |
|---|---|---|---|---|
| $\nu(CO)$, cm$^{-1}$ | 2143 | 2120–1850 | 1850–1750 | 1750–1600 |

Bridging CO ligands become less common on descending a transition metal sub-group. The increased length of the M-M bonds makes it more difficult for the CO ligand to interact effectively with both metals. However, bridging CO ligands become increasingly common as the negative charge on a cluster is increased, due to the superior ability of bridging carbonyls to delocalise electron density away from the metal centres.

Semi-bridging carbonyl ligands are also known and represent an intermediate state between terminal and bridging CO ligands. Their appearance suggests a mechanism by which carbonyl ligands can migrate over a cluster, a phenomenon that is indeed very common in cluster chemistry. This point will be explored in more depth in Chapter 5.

### 4.1.2 Other effects

Numerous factors affect the strength of the M–C and C–O bonds. The position of the metal in the periodic table plays a role, as 4d and 5d orbitals are larger than the 3d orbitals and overlap more effectively with the $\pi^*$ CO antibonding orbitals. Second and third row transition metals therefore form stronger M–CO bonds than those of the first row transition metals.

The charge on a cluster has a marked influence on the metal-carbonyl interaction. The higher the negative charge, the more strongly electron density is distributed into the CO $\pi^*$ antibonding orbitals, reducing the C–O stretching frequency. For example, oxidising or reducing the large osmium cluster $[Os_{20}(CO)_{40}]^{2-}$ has a clear effect on the $\nu(CO)$ values, as shown in Table 4.2.

TABLE 4.2

| | $[Os_{20}(CO)_{40}]$ | $[Os_{20}(CO)_{40}]^-$ | $[Os_{20}(CO)_{40}]^{2-}$ | $[Os_{20}(CO)_{40}]^{3-}$ | $[Os_{20}(CO)_{40}]^{4-}$ |
|---|---|---|---|---|---|
| $\nu(CO)$, cm$^{-1}$ | 2068, 2031 | 2059, 2019 | 2045, 2002 | 2033, 1988 | 2023, 1978 |

The majority of ligands are better $\sigma$ donors and poorer $\pi$ acceptors than CO, so substitution of CO by, for example, a phosphine affects the remaining CO ligands. Phosphines donate electron density to the metal *via* their $\sigma$ bond but do not compete effectively with CO for the available $\pi$ electron density, which increases $\pi^*$ backbonding and destabilises the CO bond. Table 4.3 shows how the average $\nu(CO)$ value decreases as the CO ligands of $Ru_3(CO)_{12}$ are substituted with $PPh_3$.

TABLE 4.3

| | $Ru_3(CO)_{12}$ | $Ru_3(CO)_{11}(PPh_3)$ | $Ru_3(CO)_{10}(PPh_3)_2$ | $Ru_3(CO)_9(PPh_3)_3$ |
|---|---|---|---|---|
| $\nu(CO)$, cm$^{-1}$ | 2042 | 2023 | 2007 | 1985 |

The value given is a weighted average of all the $\nu(CO)$ bands for each cluster.

## 4.2 LIGANDS RELATED TO CO

Nitric oxide, NO, is qualitatively similar to CO except that it contains an extra electron, which occupies a $\pi^*$ orbital. As such, it is formally a three electron donor, but bonds to metals in an otherwise analogous fashion to CO. NO can also bond to mononuclear complexes in a bent configuration as a 1e donor, but this coordination mode is very rare in clusters. Like carbonyl ligands, nitrosyl ligands can adopt bridging ($\mu$-NO) and face-capping ($\mu_3$-NO) modes, as shown in Figure 4.3 for the mixed-metal cluster $MnFe_2(\mu$-CO$)_2(\eta$-$C_5H_5)(\eta$-$C_5H_4Me)(\mu$-NO$)(\mu_3$-NO$)$. Nitrosyl ligands can also adopt more unusual bonding modes, such as the $\mu_4$–$\eta^2$ mode, in which both atoms bond over a cluster square face.

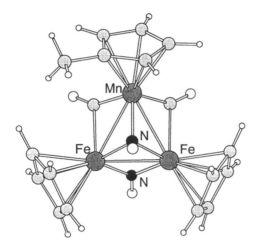

FIGURE 4.3 THE STRUCTURE OF $MnFe_2(\mu$-CO$)_2(\eta$-$C_5H_5)(\eta$-$C_5H_4Me)(\mu$-NO$)(\mu_3$-NO$)$ SHOWING EDGE-BRIDGING AND FACE-CAPPING NO LIGANDS.

The IR stretching frequency of N–O is typically in the range 1800–1900 cm$^{-1}$ for terminal NO ligands. The $\nu$(CO) values decrease with the extent of the bridging and in $MnFe_2(\mu$-CO$)_2(\eta$-$C_5H_5)(\eta$-$C_5H_4Me)(\mu$-NO$)(\mu_3$-NO$)$, the doubly and triply bridging NO ligands give rise to stretches at 1525 and 1335 cm$^{-1}$, respectively.

The isocyanide ligand, CNR, is a two-electron donor that is isostructural with CO. It is a better $\sigma$-donor and a poorer $\pi$-acceptor. Isocyanides display intense $\nu$(CN) bands in the region 2000–2190 cm$^{-1}$. A variety of bonding modes of CNR ligands are known that are unique to clusters, most of which are analogous to the various known binding modes of CO. Nitriles (NCR), also donate two electrons but bond only weakly to clusters and this property is used to good effect in synthesis (see Chapters 6 and 7).

## 4.3 PHOSPHORUS LIGANDS

Phosphine ligands (PR$_3$) are good $\sigma$-donors and poor $\pi$-acceptors, and are only found in terminal coordination sites. There are many examples of transition metal clusters with phosphine ligands, since CO is readily substituted by these two-electron ligands. Their ability to back-donate to the metal varies considerably, phosphines with electronegative substituents such as $P(OR)_3$ ) actually called phosphites) or $PF_3$ being the best π-acceptors. Whether it is the phosphorus dπ orbitals or the P–R σ* orbitals that accept the metal dπ electrons is a subject of debate, but the weight of evidence backs the latter.

When CO is substituted by a phosphine on a cluster, the coordination site adopted is restricted by the steric and electronic requirements of the comparatively bulky phosphine ligand. The progressive substitution of CO by $PR_3$ on the trinuclear cluster $Ru_3(CO)_{12}$ provides an instructive example (see Figure 4.4).

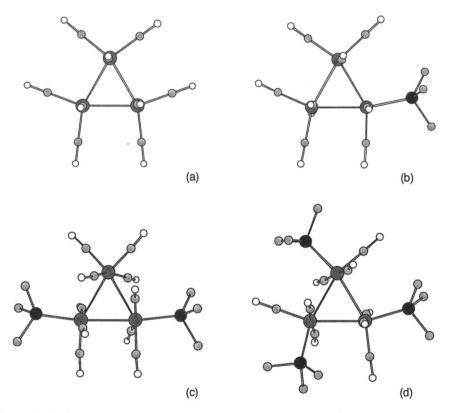

FIGURE 4.4   A COMPARISON OF THE STRUCTURES OF (a) $Ru_3(CO)_{12}$, (b) $Ru_3(CO)_{11}(PPh_3)$, (c) $Ru_3(CO)_{10}(PPh_3)_2$ AND (d) $Ru_3(CO)_9(PPh_3)_3$; FOR CLARITY PHENYL RINGS HAVE BEEN REDUCED TO A SINGLE ATOM.

There are two types of CO ligand in $Ru_3(CO)_{12}$. The ones in the same plane as the $Ru_3$ triangle are referred to as *equatorial*, those perpendicular to the $Ru_3$ triangle are *axial*. The first phosphine replaces an equatorial CO ligand and there are two reasons for this:

i.     Equatorial sites are less sterically congested, the M–M–L angle being close to 120° (cf. 90° for the axial site).

ii.   Phosphines compete poorly for $\pi$-electron density when *trans* to a carbonyl ligand (CO being a better $\pi$-acceptor) and so the phosphine prefers to occupy a coordination site *trans* to a metal–metal bond.

The next two phosphines in the triruthenium cluster also occupy equatorial positions and illustrate the principle that multi-substitution usually takes place at different metal centres.

Polydentate phosphine ligands are well known. The most common are bidentate phosphines linked by units such as $-CH_2-$, $-NR-$, $-(CH_2)_n-$, $-C_6H_4-$, $-CH=CH-$, $-C\equiv C-$ and $-(C_5H_4)Fe(C_5H_4)-$. These polydentate ligands can adopt various bonding modes and the actual bonding is strongly influenced by the geometry of the linking unit. When the backbone is flexible, the electronic requirements of the cluster can also dictate the way in which the ligand coordinates. Bidentate phosphines can chelate a single metal atom in the cluster, bridge across a metal–metal bond, or form an intermolecular link between clusters. For example, *cis-bis*-(diphenyl-phosphino)ethene $(Ph_2PCH=CHPPh_2)$ prefers to form a five-membered chelate ring with the metal (see Figure 4.5a) whereas $Ph_2PCH_2PPh_2$ prefers to bridge a metal–metal edge, also forming an unstrained five-membered ring (see Figure 4.5b). However, the stereochemistry and rigidity of $Ph_2P(C_6H_4\text{-}1,4)PPh_2$ means that it is unable to form two bonds to the same cluster and instead bonds to two separate clusters, as shown in Figure 4.5c. Interestingly, two isomers of $H_4Ru_4(CO)_{10}(Ph_2PCH_2CH_2PPh_2)$ are known, one with a chelating ligand and one with a bridging ligand.

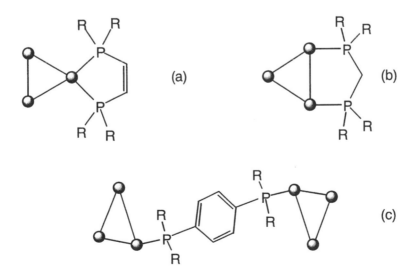

FIGURE 4.5   EXAMPLES OF BIDENTATE PHOSPHINE LIGAND COORDINATION; (a) CHELATING A METAL CENTRE, (b) BRIDGING A METAL–METAL BOND AND (c) BRIDGING TWO CLUSTERS.

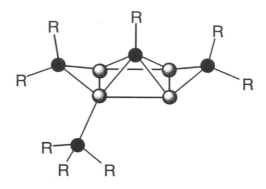

FIGURE 4.6   THE STRUCTURE OF $Ru_4(\mu\text{-}H)_4(\mu_4\text{-}PPh)(\mu_2\text{-}PPh_2)_2(PPh_3)(CO)_7$.

Loss of one of the R-groups from a coordinated phosphine results in a phosphido ligand, $\mu\text{-}PR_2$, a 3-electron donor ligand that bridges a metal–metal bond. Loss of a second group affords a phosphinidene ligand, PR, a 4-electron donor ligand that is found in face-capping positions ($\mu_3\text{-}PR$ or $\mu_4\text{-}PR$). Loss of all R-groups gives a bare phosphorus atom and these are located in interstitial sites (see Section 4.5). An example of a cluster with three different types of phosphorus ligand is the tetranuclear cluster $Ru_4(\mu\text{-}H)_4(\mu_4\text{-}PPh)(\mu_2\text{-}PPh_2)_2(PPh_3)(CO)_7$, depicted in Figure 4.6.

Phosphido ligands are often formed by inducing CO loss from a cluster to which a secondary phosphine ($PR_2H$) is coordinated. The hydrogen migrates to the cluster and the $PR_2$ group adopts a bridging position, resulting in a phosphido ligand and satisfying the electron count of the cluster. Similarly, loss of two CO ligands from a cluster with a primary phosphine, $PRH_2$ can result in the formation of two hydride ligands and a face-capping phosphinidene group (see Scheme 4.1).

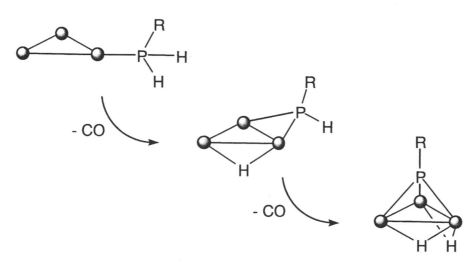

SCHEME 4.1   THE STEPWISE CONVERSION OF A PRIMARY PHOSPHINE ($PRH_2$) TO A PHOSPHIDO ($\mu\text{-}PRH$) AND THEN A PHOSPHINIDENE ($\mu_3\text{-}PR$) LIGAND.

The nitrogen analogues of phosphines, $NR_3$, are much poorer ligands because they have no orbitals of suitable energy to accept electron density from the metal, and act essentially as donor ligands. Arsenic and antimony, the heavier congeners of group 15, display coordination chemistry similar to that of phosphorus but have been much less studied.

## 4.4 INTERSTITIAL MAIN GROUP ATOMS

Polyhedral clusters contain cavities, the size and shape of which is dependent on the geometry of the metal core and the interatomic separation. It is possible for main group atoms to occupy a cavity, and this internal atom is known as an *interstitial atom*. A partially encapsulated main group atom is termed *semi-interstitial*, and in certain cases is sufficiently exposed to undergo reaction. Examples of both types are shown in Figure 4.7.

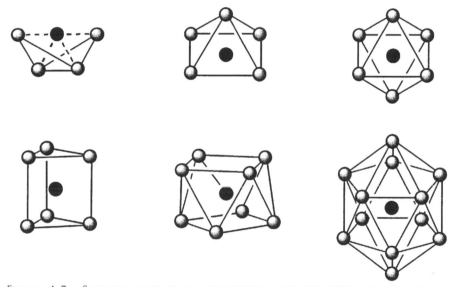

FIGURE 4.7    SOME EXAMPLES OF SEMI-INTERSTITIAL AND INTERSTITIAL ATOMS IN VARIOUS CLUSTERS.

The radius of the interstitial atom influences the type of cavity that it can occupy (only hydrogen, for example, is known to fit inside a tetrahedral cavity). Theoretically, the geometry of a cavity defines the size of the atom that can be encapsulated, based on the radii of the metal atoms ($r_{metal}$) to that of the interstitial atom ($r_{interstitial}$). Figure 4.8 shows the relevant radii for an octahedral cavity, and for comparison purposes, a table of cavity sizes of some other common cluster geometries is given.

However, the metal framework and the interstitial atom tend to be 'soft' enough so that a less than perfect fit can be accommodated. Light main group elements such as boron, carbon and nitrogen are particularly flexible in this regard. For example, an octahedral cavity should only be able to contain atoms with $r_{int}/r_{met} \leq 0.414$, but B, C and N are commonly found in octahedral cavities despite a radius ratio of about 0.6.

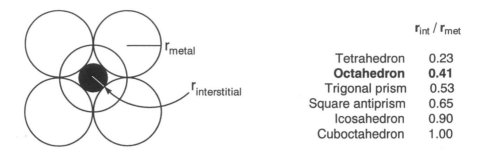

| | $r_{int} / r_{met}$ |
|---|---|
| Tetrahedron | 0.23 |
| **Octahedron** | **0.41** |
| Trigonal prism | 0.53 |
| Square antiprism | 0.65 |
| Icosahedron | 0.90 |
| Cuboctahedron | 1.00 |

FIGURE 4.8 THE INTERSTITIAL CAVITY OF AN OCTAHEDRAL CLUSTER, SHOWING $R_{MET}$ AND $R_{INT}$, WITH A LIST OF CAVITY SIZES FOR CLUSTERS OF DIFFERENT GEOMETRIES.

Second row p-block atoms are too large to fit into octahedral or smaller cavities, and third row elements can only occupy square antiprismatic or larger cages. Even larger main group atoms have only been found in icosahedral cavities, such as the tin atom in $[Ni_{12}Sn(CO)_{22}]$. Completely encapsulated transition metal atoms can only reside in a cluster with 12 or more surface atoms. Such clusters often have cuboctahedral geometries and closely resemble a fragment of a close-packed metal lattice, as in $[Rh_{12}Pt(CO)_{24}]^{4-}$. In this cluster the platinum atom occupies the cavity defined by the 12 rhodium atoms, as shown in Figure 4.9.

In cases where the interstitial atom and the surrounding elements are the same, the radial metal–metal bond lengths are always the shortest in the cluster. This observation is consistent with the idea that the effective radius of an interstitial atom is smaller than its covalent radius.

Examples of large clusters with more than one interstitial atom are known, and these can display a remarkable resemblance to the bulk phase of certain nitrides or carbides. For example, the hexagonal prismatic cluster $[Co_{14}(\mu_6\text{-}N)_3(CO)_{26}]^{3-}$

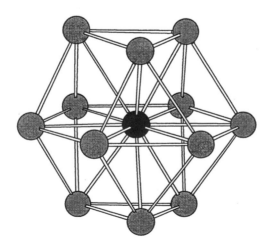

FIGURE 4.9 THE STRUCTURE OF THE METAL CORE OF $[Rh_{12}Pt(CO)_{24}]^{4-}$.

contains three nitride atoms and represents a well-defined fragment of the hexagonal packing lattice adopted by some early transition metal carbides and nitrides (such as tungsten carbide). Clusters containing polycarbide units are also known that, for example, encapsulate $C_2$ units (see Figure 4.10).

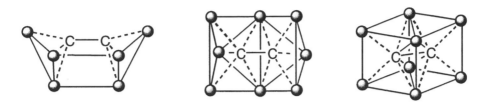

FIGURE 4.10 SOME EXAMPLES OF CLUSTERS WITH INTERSTITIAL $C_2$ UNITS.

## 4.5 HYDRIDE LIGANDS

Hydrogen atoms are common ligands in cluster chemistry and can bond to clusters in a variety of ways. Hydride ligands can bond in terminal (H), bridging ($\mu$-H, the most common type) and face-capping ($\mu_3$-H) modes. They can also occupy interstitial coordination sites (e.g. tetrahedral $\mu_4$-H or octahedral $\mu_6$-H). These last four modes are shown in Figure 4.11.

The M–H bond in terminally attached hydride ligands can be considered to be a localised two centre/two electron interaction, but when the hydride interacts with more than one metal a delocalised description is required to account for the bonding. A three centre/two electron description accounts for edge-bridging hydride ligands and fits with the experimental observation that hydride-bridged M-M bonds are slightly elongated. In fact, this phenomenon is sometimes employed to locate hydride ligands in X-ray structures from the heavy atom positions (hydrogen, with its single electron, has a negligible effect on the scattering of X-rays and is hence almost invisible). Hydrides also have a stereochemical effect on neighbouring ligands, though in clusters the steric requirements of other, bulkier ligands can overwhelm the hydrides to the extent that their 'coordination site' can not be determined. Definitive evidence for the location of hydrides is best determined by neutron diffraction (see Chapter 5).

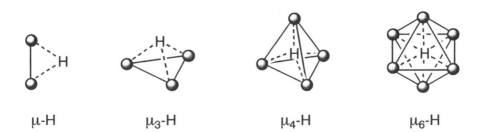

$\mu$-H $\qquad$ $\mu_3$-H $\qquad$ $\mu_4$-H $\qquad$ $\mu_6$-H

FIGURE 4.11 SOME HYDRIDE BONDING MODES.

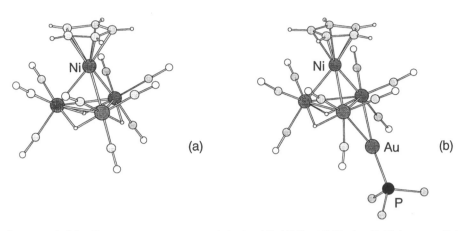

FIGURE 4.12 THE STRUCTURES OF (a) $(\mu\text{-}H)_3NiOs_3(CO)_9(\eta\text{-}C_5H_5)$ AND (b) $(\mu\text{-}H)_2NiOs_3(CO)_9(\eta\text{-}C_5H_5)(\mu\text{-}AuPPh_3)$. NOTE THE PHENYL RINGS OMITTED FOR CLARITY.

The fragment $[AuPR_3]^+$ is isolobal with $[H]^+$ and typically occupies the same bonding site as a bridging hydride ligand. For example, the mixed-metal cluster $(\mu\text{-}H)_3NiOs_3(CO)_9(\eta\text{-}C_5H_5)$ has an analogous structure to $(\mu\text{-}H)_2NiOs_3(CO)_9\text{-}(\eta\text{-}C_5H_5)(\mu\text{-}AuPPh_3)$ (see Figure 4.12).

## 4.6 UNSATURATED ORGANIC LIGANDS

### 4.6.1 Alkene and alkyne ligands

The Dewar-Chatt-Duncanson model describes the binding of alkenes to transition metals. Like the synergistic bonding of the CO ligand to a metal centre, the alkene is a two electron donor and the bonding consists of two components, as shown in Figure 4.13, via:

i.   Overlap of the $\pi$-orbital electron density with an empty $\sigma$-acceptor orbital on the metal atom.
ii.  Back-bonding from filled metal orbitals into the antibonding orbitals on the ligand.

**σ bond**

**π back-bond**

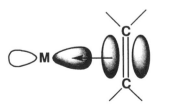
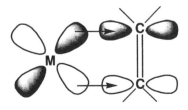

FIGURE 4.13 THE TWO COMPONENTS OF THE METAL–ALKENE BOND.

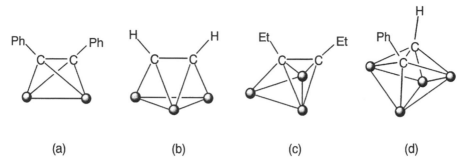

FIGURE 4.14 EXAMPLES OF COMMON ALKYNE BONDING MODES (a) $Co_2(CO)_6(\mu-C_2Ph_2)$, (b) $Os_3(CO)_{10}(\mu_3-C_2H_2)$, (c) $Fe_3(CO)_9(\mu_3-C_2Et_2)$, (d) $Co_4(CO)_{10}(\mu_4-C_2PhH)$.

The C–C bond is lengthened as a result, and the R-groups of the alkene bend away from the metal. In extreme cases where the R-groups comprise strongly electron withdrawing substituents such as $CF_3$ groups, the bonding is may be viewed as involving two 2c/2e M–C bonds and a C–C single bond. With this representation the molecule can be described as a metallocyclopropane. In clusters, the bonding of alkenes can be much more complicated, and a wide range of bonding modes for alkene ligands is known, especially for polyalkenes. Chapter 8 contains many examples of reactions involving unsaturated organic groups.

Alkynes bond to single metal centres in a similar fashion to alkenes, using a single set of their π-orbitals. However, there is a large range of bonding modes that alkynes adopt with clusters, as both π-bonds (which are at right angles to each other) may simultaneously interact with different metal centres. Figure 4.14 shows four common ways in which alkynes bond to clusters.

Due to the back donation of electron density from the metal into the $\pi^*$ orbitals of the ligand, the geometry of the alkyne is significantly distorted upon coordination to a cluster. The C–C bond of the alkyne is weakened and the substituent groups are bent away from the metals. Acetylenic protons (RCCH) are often appreciably acidic and can react with clusters to form hydride complexes (see Chapter 8). The resulting bonding modes involve combinations of σ and π bonding and are often complicated.

### 4.6.2 Unsaturated rings

Cyclopentadienyl ligands $[C_5R_5]^-$ are encountered throughout organometallic and cluster chemistry. In the majority of examples the bonding in clusters is identical to that in conventional mononuclear organometallic compounds; that is, the $\eta^5$ mode in which all five carbon atoms π-bond to a single metal centre as a five electron donor. Cyclopentadienyl ligands rarely undergo reaction themselves and often simply act only to occupy coordination sites, although $\eta^5 \rightarrow \eta^3 \rightarrow \eta^5$ slippage has been proposed as an important step in many substitution reactions. Perhaps the most interesting property of cyclopentadienyl ligands is their ability to sterically enforce a different geometry to that predicted electronically. For example, the clusters $Ni_6(\eta-C_5H_5)_6$ (not a carbonyl cluster of course, but an instructive example nonetheless)

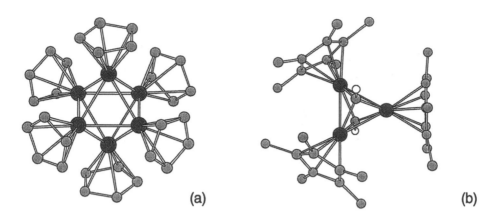

FIGURE 4.15 THE STRUCTURES OF (a) $Ni_6(\eta-C_5H_5)_6$ AND (b) $Rh_3(\mu_3-CO)_2(\eta-C_5Me_5)_3$.

and $M_3(\mu_3-CO)_2(\eta-C_5Me_5)_3$ (M = Co, Rh) (see Figure 4.15) have different electron counts to those expected.

$Ni_6(\eta-C_5H_5)_6$ has four electrons in excess of that predicted by PSEPT for an octahedral cluster (see Chapter 2), and is an example of a cluster in which the steric demands of the ligands have exceeded the electronic requirements of the metal core. On the other hand, the $M_3(\mu_3-CO)_2(\eta-C_5Me_5)_3$ clusters have two fewer electrons than expected for a triangular cluster, the bulkiness of $(\eta-C_5Me_5)$ making the coordination of another ligand sterically unfavourable.

The ability of cyclopentadienyl ligands to stabilise favoured geometries despite unusual electron counts enables some clusters to act as 'electron reservoirs'. Metal clusters generally have a greater number of accessible redox states than mononuclear complexes and can often easily and reversibly change oxidation states. For example, the tetrahedral iron cluster $Fe_4(\mu_3-CO)_4(\eta-C_5H_5)_4$ shown in Figure 4.16 remains intact in four overall charge states, +2, +1, 0 and −1.

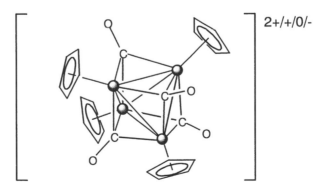

FIGURE 4.16 THE STRUCTURE OF $Fe_4(\mu_3-CO)_4(\eta-C_5H_5)_4$.

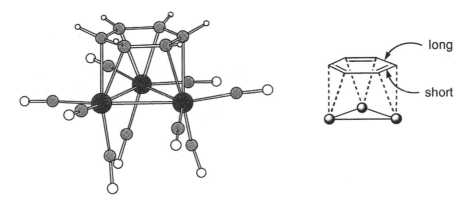

FIGURE 4.17 THE STRUCTURE OF $Ru_3(CO)_9(\mu_3\text{-}\eta^2{:}\eta^2{:}\eta^2\text{-}C_6H_6)$ AND A DIAGRAM EMPHASISING THE KEKULÉ-TYPE BENZENE RING.

Clusters without cyclopentadienyl ligands can also display a wide range of oxidation states, but examples usually involve much larger clusters where the metal core can 'breathe' to accommodate the changes in charge (see Chapter 7).

Benzene and related arenes generally bond to clusters in one of two modes. In both cases the ligand provides six electrons to cluster bonding. One of these modes, $\eta^6$-coordination to a single metal atom, is familiar from mononuclear compounds such as $Cr(CO)_3(\eta\text{-}C_6H_6)$. The other bonding mode involves $\mu_3$-coordination of the $C_6$ ring over a triangular cluster face. This face-capping bonding of benzene closely models the way in which benzene chemisorbs to a metal surface. The $\mu_3$-ring lies symmetrically over the trimetal face and the interaction is shown in Figure 4.17. The ring is distorted with the shorter unsaturated bonds sitting directly over the metal atoms and the longer single bonds sit directly over the metal atoms and the longer single bonds sit above the metal–metal edges. As such, the ring is more like a cyclohexa-1,3,5-triene (i.e. Kekulé benzene). In addition, the hydrogen atoms lie out of the plane of the ring, bending away from the cluster core.

Cyclic polyenes such as cyclohexadiene, cycloheptatriene and cyclooctatetraene exhibit a variety of binding modes in which the ligand can bind facially or apically using one, some, or all of their double bonds. Two examples of clusters with cyclic polyene ligands are presented in Figure 4.18, $Co_4(CO)_6(\mu_3\text{-}\eta\text{-}C_7H_7)(\eta^5\text{-}C_7H_9)$ and $Co_4(CO)_6(\mu_3\text{-}\eta\text{-}C_8H_8)(\eta^4\text{-}C_6H_8)$.

Many other examples are known in which cyclic polyenes bind to clusters. As a general rule polyenes donate the same number of electrons as their hapticity, so the examples above have polyene ligands that donate $7 + 5 = 12$ and $8 + 4 = 12$ electrons respectively. Both clusters have the usual count of 60 valence electrons for a tetrahedral cluster.

### 4.7 OTHER LIGANDS

Despite the importance of alkyl ligands ($CR_3$) in mononuclear organometallic chemistry, they are extremely rare in clusters. Bridging alkylidene ligands ($\mu\text{-}CR_2$)

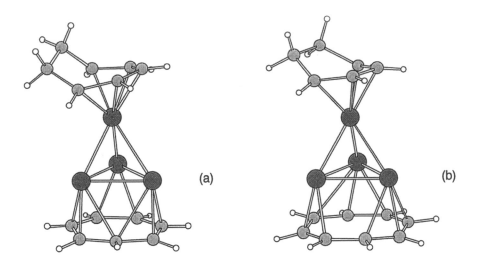

FIGURE 4.18 THE STRUCTURES OF (a) $Co_4(CO)_6(\mu_3\text{-}\eta\text{-}C_7H_7)(\eta^5\text{-}C_7H_9)$ AND (b) $Co_4(CO)_6(\mu_3\text{-}\eta\text{-}C_8H_8)(\eta^4\text{-}C_6H_8)$.

are more common and the face-capping alkylidyne ($\mu_3$-CR) ligand is probably the best studied $\sigma$-bonded carbon ligand. Silicon, germanium and tin show similarities to carbon in the ways in which they bond to clusters. As well as the simple terminal ($ER_3$, 1e) mode, bridging ($\mu$-$ER_2$, 2e) and face-capping ($\mu_3$-ER, $\mu_4$-ER or $\mu_5$-ER, 3e) modes are also known (see Figure 4.19).

When the Group 14 atom is bonded solely to metals it can be regarded as 'semi-interstitial'. However, since there are always more E–M bonds than M–M bonds, the Group 14 atom is contributing far more to the structure than simply occupying a cavity like genuine interstitial atoms. Lead is less important than the other Group 14 elements in cluster chemistry since it forms relatively long E–M bonds. This characteristic often results in the formation of 'open' compounds such as $Pb[Co(CO)_4]_4$, which has no Co–Co bonds.

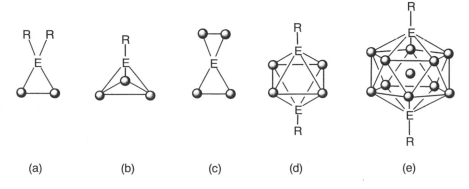

FIGURE 4.19 SOME EXAMPLES OF GROUP 14 LIGANDS BONDING TO CLUSTERS (a) $H_2Re_2(CO)_8(\mu\text{-}SiPh_2)$, (b) $Co_3(CO)_9(\mu_3\text{-}CPh)$, (c) $\{Fe_2(CO)_8\}_2(\mu:\mu\text{-}Si)$, (d) $Co_4(CO)_{11}(\mu_4\text{-}GeMe)_2$, (e) $[Ni_{11}(CO)_{18}(\mu_5\text{-}SnMe)_2]^{2-}$.

Group 16 containing compounds and the elements themselves (O, S, Se and Te) are versatile ligands in carbonyl cluster chemistry. They are found in various bonding modes in which they can contribute two to four electrons to the cluster, as shown in Figure 4.20.

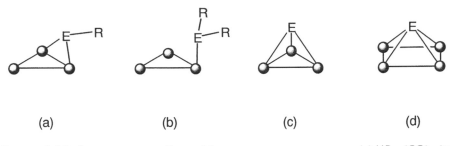

(a)          (b)          (c)          (d)

FIGURE 4.20 SOME EXAMPLES OF GROUP 16 LIGANDS BONDING TO CLUSTERS (a) $HRu_3(CO)_{10}(\mu$-SEt), (b) $Os_3(CO)_{11}(SMe_2)$, (c) $RuCo_2(CO)_9(\mu_3$-S), (d) $Co_4(CO)_{10}(\mu_4$-S)$_2$.

Group 16 ER ligands usually bridge a metal–metal bond ($\mu$-ER), contributing three electrons to the cluster. $ER_2$ ligands donate a lone pair of electrons. The elements themselves typically cap triangular or square metal faces ($\mu_3$-E, $\mu_4$-E) and in both cases donate four electrons towards cluster bonding. Figure 4.21 shows the structure of the unusual mixed-metal, mixed-ligand cluster $Fe_2Mo_2(CO)_6(\eta$-$C_5H_5)_2$ $(\mu_3$-S)($\mu_3$-Se)($\mu_4$-Te), which displays three different chalcogens in two common binding modes.

The halogens (F, Cl, Br, I) are reasonably well known as ligands in carbonyl clusters and donate one electron when bound terminally, three electrons when bridging a metal–metal bond and five electrons when capping a face.

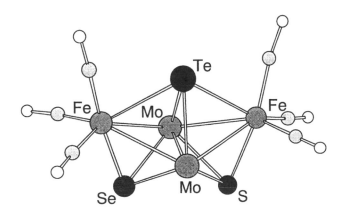

FIGURE 4.21 THE STRUCTURE OF $Fe_2Mo_2(CO)_6(\eta$-$C_5H_5)_2(\mu_3$-S)($\mu_3$-Se)($\mu_4$-Te); FOR CLARITY THE $C_5H_5$ LIGANDS ARE NOT INCLUDED.

# CHAPTER 5

# CHARACTERISATION TECHNIQUES

In the early days of transition metal carbonyl cluster chemistry, only a handful of analytical and spectroscopic techniques were available suitable for their characterisation. In fact, definite characterisation of the first metal–metal bond in a molecular compound took several years to complete. Fortunately, the development of new techniques combined with massive improvements in computing power has now made the characterisation of new cluster compounds routine. Chemists today can choose from a wide range of techniques that help them study novel compounds and materials. Four techniques are of particular importance to the study of transition metal carbonyl clusters, these being infrared spectroscopy, mass spectrometry, NMR spectroscopy and X-ray crystallography. Other techniques are generally only employed where these methods fail to provide all of the information required to characterise the cluster. Peculiarities of each technique in their application to cluster chemistry is described and a brief summary of some of the less frequently used characterisation methods is given.

## 5.1    INFRARED SPECTROSCOPY

Infrared (IR) spectroscopy is fast, sensitive and ideally suited to the study of transition metal carbonyl compounds. The $v(CO)$ stretching frequencies are generally sharp and strong, and appear in the range 2150 and 1550 cm$^{-1}$. Because few other bonds absorb in this region, IR study of the $v(CO)$ bands may even be performed *in situ*, in a wide range of solvents without interference. The broad range of stretching frequencies is due to the fact that the bonding of CO to a cluster is affected not only by the charge of the cluster, the metal to which it is attached and the presence of other ligands, but also by the bonding mode which the CO adopts (see Chapter 4).

In mononuclear complexes the number of bands in an IR spectrum is easily predicted from symmetry rules, but this approach works only for the smallest clusters. Despite the increasing number of CO ligands and the greater complexity of larger clusters, the IR spectra of these compounds are often very simple and hence contain little structural information.

IR spectroscopy is also useful for detecting metal–hydrides. The M-H stretch appears between 2200 and 1600 cm$^{-1}$ for terminal hydrides and from 1600 to 800 cm$^{-1}$ for bridging hydrides. However, the peaks are often broad and weak and hidden under the $v(CO)$ bands, so $^1$H NMR is usually a superior technique for the identification of hydride ligands.

Probably the greatest utility of IR spectroscopy in cluster chemistry is in following the progress of a reaction. Because only very small quantities of material are needed and it is a rapid, straightforward technique, IR spectroscopy can be used to examine aliquots of a reaction solution and thus determine the point at which the reaction is complete, and perhaps identify intermediates. Recently, optical fibres connected to IR spectrometers have been placed directly into reaction vessels and used to continuously monitor the reaction taking place.

## 5.2    MASS SPECTROMETRY

Mass spectrometry involves the production of gas-phase ions and their separation and detection according to their mass/charge (*m/z*) ratio. There are a number of methods used to produce the ions (ionisation techniques) and to separate them (all of which use electric fields, sometimes in conjunction with magnetic fields). Separation of the ions is outside the scope of this book, but how the ions are generated has a significant effect on the type of information provided, and will be briefly discussed.

*Electron impact* (EI) is the classical method used to generate ions. It involves heating a sample under high vacuum and exposing it to high energy electrons, which results in both ionisation and fragmentation. It is suitable for neutral, thermally stable and reasonably volatile compounds only (large, highly substituted or ionic clusters are not suitable), and extensive fragmentation may mean that the molecular ion, $[M]^+$, is not detected. *Fast atom bombardment* (FAB) of a sample results in the target molecule being blasted into the gas phase, acquiring charge and staying largely intact. It works on the principle that if a compound is suddenly superheated, vaporisation will occur before unimolecular decomposition. FAB can provide spectra of involatile and ionic clusters. A relatively new technique is *matrix assisted laser desorption ionisation* (MALDI), which works on an analogous principle to FAB, but the atoms are replaced by a laser beam and a solid matrix (or no matrix $\rightarrow$ LDI) is used. Successful analyses of a variety of cluster compounds, both ionic and neutral, have been carried out. Another new technique is *electrospray ionisation* (ESI); a solution is injected into a strong electric field, and the solvent is evaporated from the ions by a counter flow of inert gas. Ionisation is carried out chemically if the target compound is neutral. ESI is particularly suitable for ionic compounds, which are already charged. Little or no fragmentation of the ion occurs, but can be induced if structural information as well as the molecular weight is required.

The principle information that is provided by mass spectrometry is the molecular weight. However, mass spectrometry also provides information on the ion composition by the isotope pattern and structural information by examination of the fragments. Figure 5.1 show mass spectra of the high nuclearity cluster $[Os_{10}C(CO)_{24}]^{2-}$ gathered by the two of the most modern techniques, ESI and LDI.

Molecular weight information is provided by ESI, as the spectrum consists solely of a peak corresponding to the dianion at 1294 *m/z* and the molecular weight of 2588 Da is easily derived. LDI provides additional structural information, as loss of CO ligands can be observed in the spectrum. It should be noted, however, that in the LDI process an electron is added to the cluster so that the monoanion $[Os_{10}C(CO)_{24}]^-$ is actually observed.

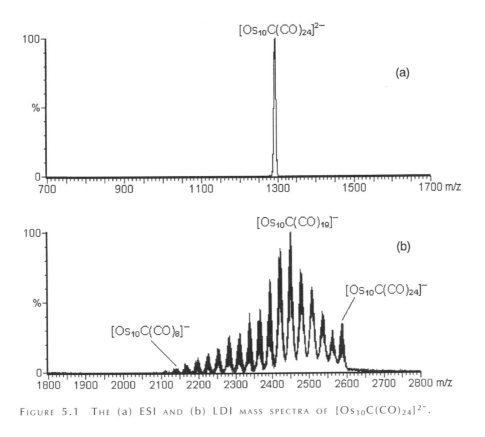

FIGURE 5.1 THE (a) ESI AND (b) LDI MASS SPECTRA OF $[Os_{10}C(CO)_{24}]^{2-}$.

## 5.3 NUCLEAR MAGNETIC RESONANCE SPECTROSCOPY

Nuclear magnetic resonance (NMR) is the most powerful *spectroscopic* technique for the determination of structure in solution. NMR provides detail on the environment of an atom (via the chemical shift, $\delta$), the relative numbers of magnetic nuclei of the same type (integration of peaks) and the presence of neighbours (coupling constants).

### 5.3.1 $^1$H NMR spectroscopy

$^1$H NMR is a very useful tool for identifying the presence of hydride ligands. Terminal, bridging and face-capping ligands have characteristically highfield signals, generally in the range $\delta$ –5 to –30. The chemical shift of interstitial hydrides can be found across an enormous range, from $\delta$ +23.2 for $[Co_6(\mu_6\text{-H})(CO)_{15}]^-$ to $\delta$ –29.3 for $[Rh_{13}(\mu_5\text{-H})_3(CO)_{24}]^{2-}$. However, coupling between inequivalent $^1$H nuclei or other nuclei such as $^{31}$P or $^{103}$Rh (both 100% abundant and spin $^1/_2$) can give additional information on the environment of the hydride ligands.

Quite often, the number of hydride ligands on a cluster is not readily determinable by the usual method of integrating peak areas of the hydride signals against, say, that of an organic ligand. Relaxation times vary greatly depending upon whether the $^1$H nuclei is bound to carbon or to a metal and in general hydride signals are much smaller than expected.

### 5.3.2 Variable-temperature NMR spectroscopy

A property of NMR spectroscopy is that the experiment has an intrinsically long timescale, in that it takes ~$10^{-4}$ seconds to 'examine' a nucleus (cf. IR, ~$10^{-13}$ seconds). This property is useful as many dynamical processes are much faster than the NMR timescale and the signals corresponding to the exchanging groups or ligands are averaged into a single peak. This phenomenon is used to advantage in variable temperature (VT) NMR, whereby a sample is cooled down to the point at which the dynamical process slows below the sampling rate. In some cases, a variety of processes may be identified, especially with respect to the scrambling of carbonyl ligands (see section 5.3.3) and the migration of hydride ligands about a metal core. A detailed analysis of the line-widths in the spectra can provide estimates of the energies involved in the fluxional process.

Organic ligands also undergo dynamic processes. For example, at –85°C the $^1$H NMR spectrum of $Os_3(CO)_8(\mu_3\text{-}C_6H_6)(\eta^2\text{-}C_2H_4)$ shows ten resonances, indicating that each proton is in a unique environment and that the ligands are all 'locked' in place. However, upon warming to room temperature, the signals of the ring ($H_1$-$H_6$) and those of the ethene ($H_A$-$H_D$) coalesce into a single resonance for each ligand (see Figure 5.2). This observation is consistent with rapid rotation of the ligands in a 'helicopter-like' motion.

### 5.3.3 $^{13}$C NMR spectroscopy

Terminal CO ligands display resonances from δ +180 to +210, whereas the chemical shift of bridging carbonyls is typically at higher frequencies. Samples usually need to be enriched with $^{13}$CO to obtain sufficiently strong signals for analysis. Carbonyl ligands on a cluster are able to undergo structural rearrangement on the NMR timescale, and this fluxional behaviour is called carbonyl scrambling. Fluxionality is a common phenomenon in organometallic chemistry, and occurs whenever a

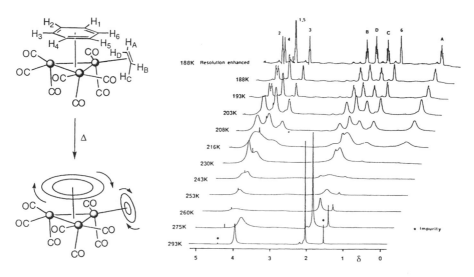

FIGURE 5.2  THE VARIABLE TEMPERATURE $^1$H NMR SPECTRA OF $Os_3(CO)_8(\mu_3\text{:}\eta^2\text{:}\eta^2\text{:}\eta^2\text{-}C_6H_6)(\eta^2\text{-}C_2H_4)$.

molecule has a number of closely related structures accessible by relatively low energy barriers. Carbonyl scrambling in clusters is a classic example of fluxionality, and is best studied using variable temperature $^{13}$C NMR spectroscopy. The relationship of the CO ligands with the internal metal unit change during scrambling, for example, from terminal to edge bridging or face bridging. Because CO donates two electrons no matter what its coordination mode, concerted migrations such as these do not require the involvement of additional ligand electrons.

### The ligand polyhedral model

Solid-state NMR spectroscopy has provided evidence for an alternative view of carbonyl scrambling. The Ligand Polyhedral Model is based on the idea that the metal core librates (rocks gently) within the polyhedron defined by the ligands of the cluster. For example, the 12 CO ligands of $Fe_3(CO)_{12}$ and $Co_4(CO)_{12}$ form a (slightly distorted) icosahedron, and the cavity in the centre contains the metal core (a triangle or tetrahedron, respectively).

Small reorientations of the metal core within the icosahedral shell of CO ligands provides a mechanism by which all of the CO ligands are rendered equivalent on the NMR timescale, despite both clusters comprising a mixture of terminal and bridging CO ligands in different chemical environments.

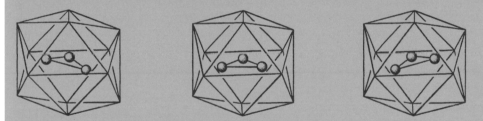

The model also provides an elegant rationalisation of the isomeric structures of simple binary carbonyl clusters. For example, the ligand polyhedron about $Ir_6(CO)_{16}$ forms a tetracapped truncated tetrahedron. One orientation of this polyhedron about the octahedral metal core results in an isomer with four face-capping CO ligands and two terminal CO ligands per metal atom. A simple rotation of the metal octahedron about the $C_4$ axis changes the $\mu_3$-CO face-capping ligands into $\mu_2$-CO edge bridging ligands, which defines the second isomeric form of $Ir_6(CO)_{16}$.

Many nuclei other than $^1$H and $^{13}$C can be studied using NMR spectroscopy. For example, the presence and environment of phosphorus ligands is conveniently established using $^{31}$P NMR spectroscopy. The $^{31}$P nucleus is 100% naturally abundant, sensitive and spin $^1/_2$, making spectra quick and easy to collect. As such, reactions with phosphines and other phosphorus containing ligands can be followed by $^{31}$P NMR spectroscopy. Other suitable nuclei such as $^{11}$B, $^{19}$F and $^{29}$Si are also found in some ligands. Metals are less easily studied, but some have abundant spin

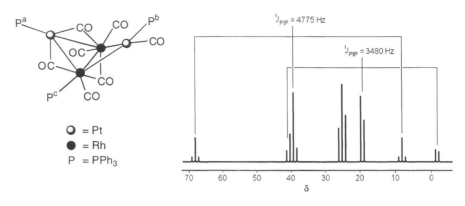

FIGURE 5.3 THE $^{31}$P-{$^1$H} SPECTRUM AND STRUCTURE OF Pt$_2$Rh$_2$(CO)$_7$(PPh$_3$)$_3$.

$^1/_2$ nuclei that provide useful information on the structure of the core and sometimes on the distribution of the ligands. Metals that have provided good spectra include $^{103}$Rh, $^{183}$W, $^{187}$Os and $^{195}$Pt. As an example of the usefulness of multinuclear NMR spectroscopy, the $^{31}$P-{$^1$H} spectrum and structure of the mixed-metal cluster Pt$_2$Rh$_2$(CO)$_7$(PPh$_3$)$_3$ are shown in Figure 5.3.

The spectrum is complicated to interpret, involving coupling between $^{31}$P, $^{103}$Rh and $^{195}$Pt nuclei, but provides much structural information. For example, the PPh$_3$ ligands bound to the platinum atoms (P$^a$ and P$^b$) display satellite peaks due to large one bond Pt-P $^1J_{Pt-P}$ coupling constants. The intensity of these peaks is reduced because the $^{195}$Pt nucleus is only 33.8% abundant. One-bond Rh-P coupling constants are much smaller than their platinum counterparts, $^1J_{Rh-P}$ = 125 Hz in this case, so no satellite peaks are seen for P$^c$.

NMR spectroscopy has been used to show that in some clusters, the metal core itself is fluxional. For example, the $^{31}$P NMR spectrum of [Rh$_9$P(CO)$_{21}$]$^{2-}$ at 40°C shows a ten line multiplet at δ 282.3 for the highly deshielded interstitial phosphorus atom, due to coupling to nine equivalent rhodium atoms. The pattern becomes more complex on cooling to −80°C, suggesting that the rhodium skeleton is fluxional. The most likely mechanism is formation of a Rh-Rh bond across the basal square face to produce a tricapped trigonal prismatic core, followed by cleavage of another Rh-Rh bond to reform the capped square antiprism (see Figure 5.4). In this way all nine rhodium atoms are rendered magnetically equivalent.

### 5.4 X-RAY CRYSTALLOGRAPHY

X-ray crystallography has contributed more than any other characterisation technique to the study of cluster chemistry. The ability of crystallography to provide detailed structural information has allowed the huge range of metal frameworks and unusual binding modes of ligands peculiar to this area of chemistry to be established.

X-ray crystallography is the most powerful method (in widespread use) for determining the structure of solids. The regularly packed molecules within a single

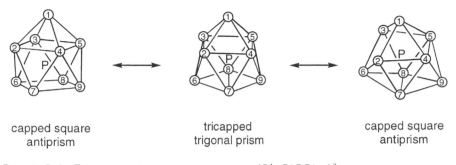

capped square
antiprism

tricapped
trigonal prism

capped square
antiprism

FIGURE 5.4 THE SKELETAL REARRANGEMENT OF $[Rh_9P(CO)_{21}]^{2-}$.

crystal act as a three-dimensional diffraction grating that scatters X-rays in a predictable way. The scattered beams can be detected and measured, and with the assistance of computers, a highly accurate model of the electron density within the crystal can be created to match the experimental observations.

### 5.4.1 Representations of X-ray structures
Pictorial representations of X-ray structures vary widely. The conventional 'ball-and-stick' type used in this book, with simple spheres of varying radius according to atomic weight, has been chosen for clarity. *Space filling* models show the van der Waals radius of each atom and are useful for showing the topography of the outer surface of a molecule. As an example, the ball-and-stick model of $Fe_5C(CO)_{15}$ suggests that the carbide atom is quite exposed, whereas the space filling model shows clearly that it is in fact well protected from attack by the surrounding ligands (see Figure 5.5).

Atoms may also be drawn as ellipsoids, as this shape better fits the real distribution of electron density of the atom. These atomic displacement ellipsoids depict the vibration of atoms, and because bends generally cause greater displacement than stretches, the ellipsoids are rarely spherical. Bridging carbonyl ligands have larger

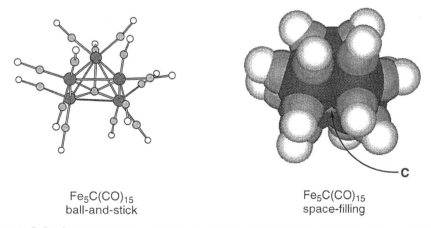

$Fe_5C(CO)_{15}$
ball-and-stick

$Fe_5C(CO)_{15}$
space-filling

FIGURE 5.5 A COMPARISON OF THE BALL-AND-STICK MODEL AND THE SPACE-FILLING MODEL OF $Fe_5C(CO)_{15}$.

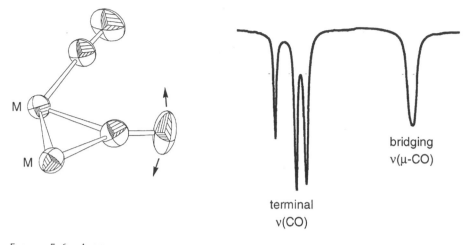

FIGURE 5.6  A COMPARISON OF THE THERMAL ELLIPSOIDS OF A TERMINAL AND A BRIDGING CARBONYL TOGETHER WITH AN IR SPECTRUM SHOWING THE DIFFERENCE IN WIDTH OF THE TWO TYPES OF STRETCHES.

ellipsoids than terminal carbonyl ligands, due to pronounced hinge-like movement of the carbonyl ligand (see Figure 5.6). This phenomenon has been cited as the reason for the fact that the $\nu(\mu\text{-CO})$ bands in the infrared spectrum are significantly broader than the terminal $\nu(CO)$ bands.

### 5.4.2 Disorder

Disorder arises from the random distribution of atoms or groups of atoms across two or more sites in a unit cell, and causes problems in crystallography. *Static* disorder results in an average structure that includes partial atoms, a classic carbonyl cluster example being the structure of $Fe_3(CO)_{12}$ in which the triangle of iron atoms is disordered over two sites in a 'Star of David' pattern inside a distorted icosahedral shell of carbonyl ligands. Figure 5.7 shows an illustration of $Fe_3(CO)_{12}$ alongside a comparison to the non-disordered structure of $Ru_3(CO)_{12}$, a triangle of ruthenium atoms inside a cubeoctahedral shell of carbonyl ligands.

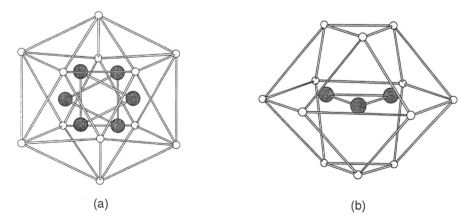

(a)  (b)

FIGURE 5.7  A COMPARISON OF (a) DISORDERED STRUCTURE OF $Fe_3(CO)_{12}$ AND (b) ORDERED STRUCTURE OF $Ru_3(CO)_{12}$.

Another type of static disorder is *substitutional* disorder, where a compound has sites, which can contain different atoms in any ratio. A striking example of bimetallic substitutional crystal disorder at only certain crystallographic sites is in the high nuclearity trimetallic carbonyl cluster $[Au_6Pd_6(Pd_{6-x}Ni_x)Ni_{20}(CO)_{44}]^{6-}$. The X-ray structure of this cluster was determined for seven different crystals and in each case, at six sites on the cluster, either nickel or palladium could be found occupying the position. The value of $x$ in $(Pd_{6-x}Ni_x)$ was found to range from 2.1 (65% Pd, 35% Ni) to 5.5 (8% Pd, 92% Ni).

*Dynamic* disorder is a slightly different phenomenon and involves the rapid rotation of part of the structure in the solid state. Collecting the structure at low temperature, which may stop the movement, can alleviate this problem. While disorder problems may not affect the part of the structure that is interesting, X-ray diffraction is an 'all-or-nothing' technique in which all components contribute to the final picture. The entire structure must be precisely modelled in order to get an accurate result.

## Microscopy

It is difficult to grow single crystals of very large clusters. Big clusters are often nearly spherical and can pack in almost any orientation, a situation not conducive to long-range ordering. The likelihood of mixtures of very similar clusters being present in solution further exacerbates the problem. However, as clusters increase in size, they enter a realm in which it is possible to directly image the metal core. Technology in this area is continually improving and with modern high resolution transmission electron microscopy (HRTEM), resolution on the scale of ~2 Å is possible. The picture below shows an HRTEM image of $[Os_{20}(CO)_{40}]^{2-}$ supported on carbon. The tetrahedral shapes of the clusters may be identified in the image.

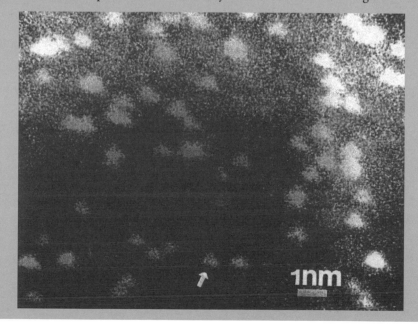

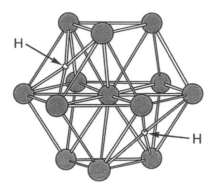

FIGURE 5.8 THE NEUTRON DIFFRACTION STRUCTURE OF $Rh_{13}H_2(CO)_{24}$.

### 5.4.3 Neutron diffraction

X-ray crystallography is poor at identifying light atoms in the presence of heavy ones, because the X-rays scatter from electrons. The heavy atoms therefore dominate the pattern and light atoms (hydrogen in particular, with its single electron) are simply lost in the 'noise'. While most hydrogen atoms appear in entirely predictable locations in organic compounds, hydride ligands on metal clusters may be found in any number of positions in and around the metal core. This problem may be addressed by careful examination of the metal-metal bond lengths, but the location of the hydride ligands is at best inferred. The most satisfactory solution is to use *neutron diffraction*, which is similar to X-ray diffraction but uses a beam of neutrons that scatter from nuclei rather than electrons. Hydrogen atoms show up clearly in the resulting map, but a nuclear accelerator is required to generate the neutron beam and large crystals are needed to compensate for the low flux of particles, and as such neutron diffraction is by no means a routine technique. Nonetheless, the results often justify the expense, as in the neutron diffraction structure of $Rh_{13}H_2(CO)_{24}$, in which the hydride ligands were shown to occupy semi-interstitial square pyramidal sites on the surface of the cubeoctahedral cluster (see Figure 5.8).

### 5.5 MISCELLANEOUS CHARACTERISATION TECHNIQUES

#### 5.5.1 Electron paramagnetic resonance spectroscopy

Electron paramagnetic (or spin) resonance (EPR or ESR) spectroscopy is similar to NMR spectroscopy in that both techniques involve the reversal of spin. However, EPR deals with unpaired electrons rather than nuclei. It is used to determine the distribution of an unpaired electron in a cluster, and has provided direct evidence for the delocalised nature of metal-metal bonding in ligated clusters. Because it deals only with paramagnetic compounds, EPR is complementary to NMR spectroscopy, but since the vast majority of carbonyl clusters are diamagnetic it is rarely encountered.

#### 5.5.2 UV-visible spectroscopy

UV-visible spectroscopy can be used to study cluster compounds, as nearly all clusters are intensely coloured. Generally, the colour of a cluster intensifies as the nuclearity increases [for example, $Os_3(CO)_{12}$ is pale yellow, $Os_5C(CO)_{15}$ is orange and $Os_6(CO)_{18}$ is black] because the HOMO-LUMO gap diminishes as the number of bonding and antibonding molecular orbitals increases. The same reasoning explains why related compounds become less highly coloured upon descending a column of the periodic table [for example, $Co_4(CO)_{12}$ is black, $Rh_4(CO)_{12}$ is red and $Ir_4(CO)_{12}$ is yellow], because the HOMO-LUMO gap increases with stronger metal-metal bonding. However, the structural information provided by UV-visible spectroscopy is rather limited, and the technique is most useful in following reactions by the disappearance and/or appearance of absorption bands. It has also been employed to confirm that species absorbed on a surface are the same as those present in solution.

### 5.5.3 Mössbauer spectroscopy

Mössbauer spectroscopy involves examination of gamma-ray resonance between a nuclear ground state and a metastable excited state. It is very sensitive to s orbital electron density, and can therefore differentiate clearly between atoms in different bonding environments. The major advantage of Mössbauer spectroscopy is its ability to probe powders, so a sample need not be soluble or form single crystals. Unfortunately, only few nuclei undergo a suitable sequence of nuclear decay processes and as far as cluster chemistry is concerned only $^{57}Fe$ and $^{197}Au$ are important.

### 5.6   CLUSTER-SURFACE ANALOGY

Metal surfaces are very important in many heterogeneous catalytic processes involving compounds that are known ligands in metal carbonyl cluster chemistry, such as CO, $H_2$ and unsaturated organic compounds. However, the actual surface chemistry that occurs at the molecular level is not fully understood. It has been proposed that discrete, molecular transition metal clusters serve as reasonable models of the metal surface. As such, clusters can accurately model small organic molecules chemisorbed to a metal surface.

The similarity between parts of a close-packed metal surface and various metal cluster frameworks is illustrated in Figure 5.9. With the knowledge gained from the way in which ligands bond to clusters (see Chapter 4), the likely binding modes of substrates to the various different surface sites such as steps, kinks, vertices, edges and faces can be visualised. The high resolution spectroscopic data obtained from molecular clusters can be compared with the low resolution data obtained directly from a metal surface. While this latter data may not be of sufficient resolution to unambiguously describe the type of bonding on the surface, comparisons with the cluster systems often helps to assign the bonding mode.

Clusters have a considerable advantage over mononuclear complexes in modelling surface reactivity in that they possess a variety of spatial arrangements of coordination sites. As well as perpendicular and opposite coordination sites, clusters also possess parallel sites on adjacent atoms, enabling substrates to bridge two (or more) sites or even coordinate at one metal centre and undergo reaction at another.

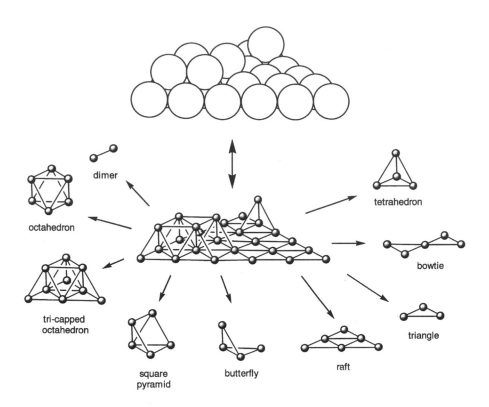

FIGURE 5.9 A COMPARISON OF A CLOSE-PACKED METAL SURFACE WITH SOME METAL CLUSTER FRAMEWORKS.

# CHAPTER 6

# SYNTHESIS

This chapter describes the ways in which transition metal carbonyl clusters may be prepared. In general, low nuclearity compounds are prepared from the reduction of metal salts whereas higher nuclearity clusters tend to be derived from lower nuclearity carbonyl compounds. Due to the vast number of carbonyl compounds known the discussion has been restricted to mostly homoleptic compounds, especially those that have been subjected to detailed physical and chemical investigations.

## 6.1  SIMPLE HOMOLEPTIC CARBONYLS

Ludwig Mond synthesised the first metal carbonyl compound, $Ni(CO)_4$, in 1890 by the direct reaction between nickel and carbon monoxide at 100°C. Within a year the same method was also used to prepare $Fe(CO)_5$, although more forcing conditions were required. It has subsequently been found that both pure nickel and iron (uncontaminated by surface oxides) will react with CO at room temperature and at atmospheric pressure to produce the appropriate carbonyl compound as shown in Scheme 6.1. The ease with which these carbonyl compounds are formed is illustrated by the fact that they are found in very low concentrations in tobacco smoke.

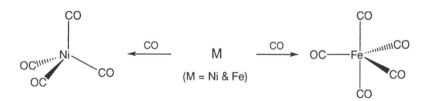

SCHEME 6.1  THE SYNTHESIS OF $Ni(CO)_4$ AND $Fe(CO)_5$ BY DIRECT REACTION OF THE METALS WITH CARBON MONOXIDE

A large number of other synthetic approaches have since been developed that yield metal carbonyl complexes for the majority of the transition metals. The main synthetic methodology involves treatment of a metal salt under a high pressure of carbon monoxide. In some cases a reducing agent must be added although it has also been found that CO can sometimes act as the reducing agent. Many ambient pressure preparations have also been developed and despite often requiring several steps, are particularly useful when facilities to conduct high-pressure reactions are not available. In addition, unstable metal carbonyl complexes and clusters have been prepared using matrix isolation techniques involving two different approaches:

i.  Trapping a stable metal carbonyl complex in an inert matrix at low temperature (typically solid argon at near absolute zero) and then photolysing to generate unstable species, e.g. $Fe(CO)_4$ and $Fe(CO)_3$ can be prepared from $Fe(CO)_5$.
ii. Condensation of a metal vapour into a low temperature matrix composed of CO itself or CO diluted in some other material. This way metal carbonyls of elements such as copper, silver and gold can be produced.

Matrix isolation has been used to prepare many unstable metal carbonyl complexes and clusters. The instability of these compounds outside of the low temperature matrix prevents them from being studied on a macroscopic scale and as such, the technique will not be discussed further here.

Both $Ru_3(CO)_{12}$ and $Os_3(CO)_{12}$ are conveniently prepared under high pressures of CO from ruthenium(III) chloride and osmium tetraoxide respectively, according to Scheme 6.2. In both reactions crystalline solid is obtained directly from the autoclave (a high pressure reaction vessel).

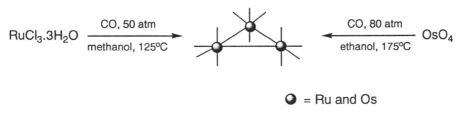

$$RuCl_3.3H_2O \xrightarrow[\text{methanol, 125°C}]{\text{CO, 50 atm}} \qquad \xleftarrow[\text{ethanol, 175°C}]{\text{CO, 80 atm}} OsO_4$$

⬤ = Ru and Os

SCHEME 6.2   THE SYNTHESIS OF $M_3(CO)_{12}$ (M = RU OR OS).

An alternative and mild route used to prepare $Ru_3(CO)_{12}$, which does not require the use of high pressures of carbon monoxide, involves bubbling CO through a refluxing solution of ruthenium(III) chloride in ethoxyethanol. Once the solution turns yellow, zinc granules and ethanol are added and the reflux is continued. The $Ru_3(CO)_{12}$ must then be purified chromatographically. Both $Ru_3(CO)_{12}$ and $Os_3(CO)_{12}$ are quite robust compounds and can be chromatographed on silica or alumina without taking precautions to exclude air and water, as can many of their derivatives.

Dicobaltoctacarbonyl is the only stable dinuclear homoleptic carbonyl of the Group 9. $Co_2(CO)_8$ may be prepared by a number of methods. One of the most reliable involves reacting cobalt carbonate with CO and $H_2$ gases in equal proportions at high pressure. The dimer is readily converted into $Co_4(CO)_{12}$ by the thermal expulsion of carbon monoxide and condensation of the unsaturated fragments either in a solid state reaction or in a solution of hexane. These reactions are illustrated in Scheme 6.3.

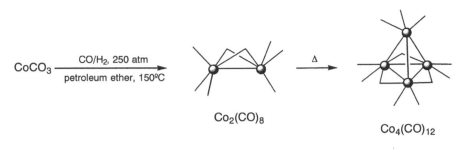

SCHEME 6.3 THE SYNTHESIS OF $Co_2(CO)_8$ AND CONVERSION TO $Co_4(CO)_{12}$.

The tetranuclear clusters $Rh_4(CO)_{12}$ and $Ir_4(CO)_{12}$ may be prepared by either high pressure reactions or at atmospheric pressure (see Scheme 6.4). The simplest route used to prepare the tetrarhodium cluster is a two step process in which the chloro-bridged dimer, $[(CO)_2RhCl]_2$, initially prepared from rhodium(III) chloride, is treated with further CO and $NaHCO_3$. $Ir_4(CO)_{12}$ is prepared in a single step from the trichloride under a high pressure of CO at elevated temperatures. Both these compounds can be handled in air, but as a solid $Rh_4(CO)_{12}$ slowly converts to $Rh_6(CO)_{16}$ over a prolonged period (in solution the process is accelerated considerably).

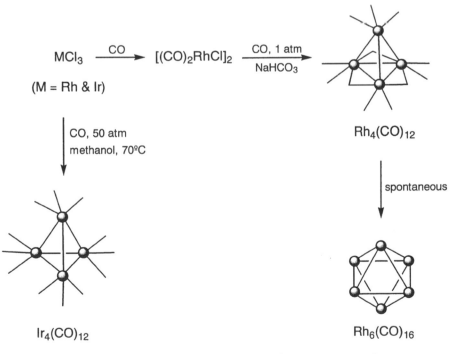

SCHEME 6.4 THE SYNTHESIS OF $M_4(CO)_{12}$ (M = Rh AND Ir) AND $Rh_6(CO)_{16}$.

The majority of clusters are prepared from smaller metal carbonyl precursors and the spontaneous conversion of $Rh_4(CO)_{12}$ to $Rh_6(CO)_{16}$ shows just how easily increases in the nuclearity of a cluster can sometimes take place. In fact, the preparation of larger clusters from smaller ones is the best strategy for the synthesis of high nuclearity clusters. Rare examples of a high nuclearity cluster prepared directly from a metal salt include the preparation of $Rh_6(CO)_{16}$ from $RhCl_3$ and the synthesis of the stacked-triangular clusters, $[Pt_3(CO)_6]_n^{2-}$ (n = 2, 3, 4 and 5) from $[PtCl_6]^{2-}$ (see below).

## 6.2 HIGH NUCLEARITY CARBONYL CLUSTERS

A considerable number of high nuclearity homoleptic carbonyl clusters, and those containing hydride ligands or interstitial atoms are known. The presence of hydride ligands and interstitial atoms, while not added deliberately, stems from the way in which many clusters are synthesised. For example, many high nuclearity clusters that are prepared in protic solvents contain hydride ligands. In addition, if the precursor cluster employed in the reaction contains hydride ligands then the product is also likely to contain hydrides. As such, this section will not be restricted to the synthesis of homoleptic, high nuclearity carbonyl clusters, as many important compounds would be excluded from the discussion.

There is a vast number of high nuclearity carbonyl clusters, made up from the elements Re, Fe, Ru, Os, Co, Rh, Ir, Ni, Pd and Pt as well as those formed from mixtures of these and other transition and main group metals. While transition metals such as Ag and Au also tend to form molecular clusters they do not form clusters with carbonyl ligands and are therefore beyond the scope of this book. These elements can be incorporated into mixed-metal clusters and some examples are given in Chapter 9. There are five main types of reactions routinely used to prepare high nuclearity carbonyl clusters:

i.    Thermolysis (heating in a solvent).
ii.   Pyrolysis (heating in the solid state).
iii.  Reductive methods.
iv.   Oxidative reactions.
v.    Redox condensation reactions.

These seemingly different synthetic methods are all similar in that the growth of the cluster requires the generation of reactive fragments that undergo condensation. In general, however, mechanistic detail on the build up of clusters is based mostly on supposition and not experimental data. In the majority of reactions in which low nuclearity carbonyl compounds are made, one product is usually formed in high yield, and it can be purified either by precipitation and washing or by crystallisation. Such simple purification methods are unsuitable for most high nuclearity cluster preparations as a number of products are usually formed. Separation of these compounds is sometimes achieved by fractional crystallisation or more commonly by column or preparative scale thin layer chromatography.

## Cluster growth mechanisms

There has been much debate concerning the mechanism for the formation of larger clusters from smaller clusters. Several mechanisms have been proposed but there is little experimental evidence available to confirm which is the most accurate. It is also possible that different processes are in operation in different reactions. However, consider the formation of a hexanuclear cluster from a trinuclear cluster. The two possibilities shown in the scheme represent different ways by which this process occurs. One mechanism (**A**) involves complete fragmentation of the cluster into mononuclear units that can then lose CO and recombine to form an octahedron. The other path (**B**) involves loss of a larger number of ligands, but without breaking any metal–metal bonds, followed by condensation of the two triangular moieties to give an octahedron.

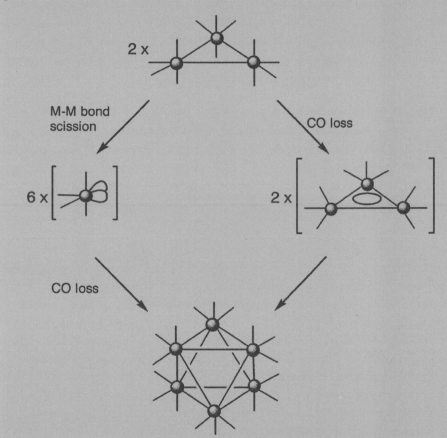

Mechanism **A** has long been the one most favoured, however, recent data obtained from gas phase studies suggest that mechanism **B** is quite feasible. Theoretical studies show that as the carbonyl ligands are removed from the intact trimer the metal–metal bonds gain a degree of unsaturated character. As such, condensation of two trimers can be viewed as a type of addition reaction.

Many clusters are stable towards oxygen and water, particularly clusters composed of second and third row transition elements. However, despite their stability, synthesis is usually carried out under an inert atmosphere. The reason that care is taken to avoid contamination with oxygen and water during the reaction is that unsaturated metal fragments are generated and it is these species that are highly reactive towards impurities. Once the reaction has been quenched and the thermodynamically favoured products are formed, the system can be opened to the atmosphere. A cautious approach is advisable as some compounds are pyrophoric (although this is relatively uncommon) or decompose slowly in air.

### 6.2.1 Preparation from salts

The synthesis of $Rh_6(CO)_{16}$ differs to that of most high nuclearity carbonyl clusters in that it can be made directly from rhodium(III) chloride (see Scheme 6.5) rather than from a lower nuclearity carbonyl precursor. The reaction is carried out under a high pressure of CO in a solution of methanol, which also acts as the reducing agent. When KOH is added (which facilitates the reduction process) then the dianion $[Rh_6C(CO)_{15}]^{2-}$ is isolated.

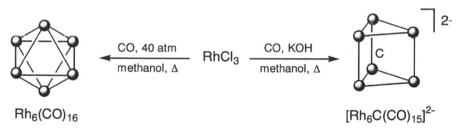

$$Rh_6(CO)_{16} \xleftarrow[\text{methanol, } \Delta]{\text{CO, 40 atm}} RhCl_3 \xrightarrow[\text{methanol, } \Delta]{\text{CO, KOH}} [Rh_6C(CO)_{15}]^{2-}$$

SCHEME 6.5   THE SYNTHESIS OF $Rh_6(CO)_{16}$ AND $[Rh_6C(CO)_{15}]^{2-}$ FROM $RhCl_3$.

The stacked triangular clusters $[Pt_3(CO)_6]_n^{2-}$ ($n = 2 - 5$), are prepared from platinum(IV) chloride. The reaction of $[PtCl_6]^{2-}$ with potassium hydroxide under an atmosphere of carbon monoxide affords a number of compounds as shown in Scheme 6.6. As the concentration of the potassium hydroxide increases the tendency to form the larger clusters increases. In fact, it is necessary to progressively reduce the platinum salt in order to produce single component products. Addition of the KOH in a single aliquot results in the formation of complicated mixtures of products. These mixtures of products are problematic, as the clusters are almost impossible to separate due to very similar solubilties. The best way to ensure that single products are obtained is to use infrared spectroscopy to follow the reaction, while slowly adding the KOH.

### 6.2.2 Thermolysis and pyrolysis

Thermolysis involves heating a low nuclearity cluster in a solvent to produce a higher nuclearity cluster. Both the solvent and the conditions have a marked influence on the products obtained. While some reactions are very specific, others are less selective and give mixtures of products. Thermolysis has been used to best effect in the preparation of high nuclearity rhenium and ruthenium clusters. For example, heating

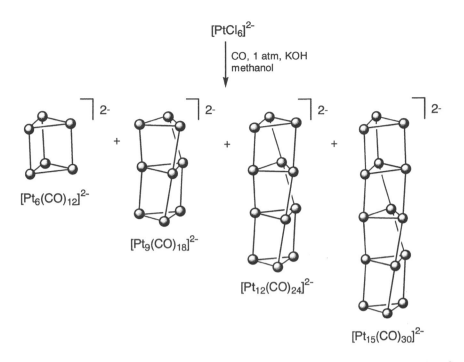

$[PtCl_6]^{2-}$

CO, 1 atm, KOH
methanol

$[Pt_6(CO)_{12}]^{2-}$ + $[Pt_9(CO)_{18}]^{2-}$ + $[Pt_{12}(CO)_{24}]^{2-}$ + $[Pt_{15}(CO)_{30}]^{2-}$

SCHEME 6.6 THE SYNTHESIS OF TRIANGULAR PLATINUM STACKED CLUSTERS, $[Pt_3(CO)_6]_n^{2-}$ (n = 2 – 5).

$[H_2Re(CO)_4]^-$ in *n*-tetradecane at 250°C results in the formation of three clusters, $[H_2Re_6C(CO)_{18}]^{2-}$, $[Re_7C(CO)_{21}]^{3-}$ and $[Re_8C(CO)_{24}]^{2-}$, illustrated in Scheme 6.7, which may be separated chromatographically. The hexanuclear cluster is only obtained in very low yield while the other two clusters are obtained in higher yields. The relative yields of the three clusters are dependent upon the precise temperature of the thermolysis and when a slightly lower temperature is used, $[H_2Re_6C(CO)_{18}]^{2-}$ is isolated as the major product. There is a clear structural relationship between the three clusters; $[H_2Re_6C(CO)_{18}]^{2-}$ has an octahedral $Re_6$ core, $[Re_7C(CO)_{21}]^{3-}$ is based on a monocapped octahedron and $[Re_8C(CO)_{24}]^{2-}$ has a *trans*-bicapped octahedral skeleton. All three clusters contain an interstitial carbide atom in the octahedral cavity.

The thermolysis of $Ru_3(CO)_{12}$ in ethanol for four hours affords the hexanuclear cluster $[HRu_6(CO)_{18}]^-$ in almost quantitative yield (see Scheme 6.8). The hydride in this cluster occupies an interstitial site and has a very distinctive $^1H$ NMR signal, *ca.* δ +20 (see Chapter 5). If the reaction time is extended to 13 hours the yield of $[HRu_6(CO)_{18}]^-$ drops and the decanuclear cluster $[H_2Ru_{10}(CO)_{25}]^{2-}$ and deposits of ruthenium metal are formed. Separation of the two clusters is achieved using chromatography. Metallic deposits are quite common in thermal reactions, because as the size of the cluster increases the ratio of CO to metal decreases and at a certain size decomposition to bare metal is thermodynamically favoured. In high boiling hydrocarbon solvents the thermolysis of $Ru_3(CO)_{12}$ affords carbide containing clusters, such as $Ru_6C(CO)_{17}$ and $[Ru_{10}C(CO)_{24}]^{2-}$, as well as a number of other

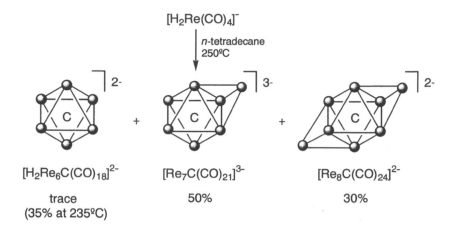

SCHEME 6.7 THE PREPARATION OF HIGH NUCLEARITY RHENIUM CLUSTERS.

products depending upon the actual solvent used. $Ru_3(CO)_{12}$ is not the only ruthenium precursor that has been used in thermolysis. For example, thermolysis of the tetranuclear species $H_4Ru_4(CO)_{12}$ affords the trianionic cluster $[HRu_{11}(CO)_{27}]^{3-}$ in high yield.

The octahedral cluster $[Co_6(CO)_{15}]^{2-}$ may be prepared in high yield from the thermolysis of $Co_2(CO)_8$ in ethanol as shown in Scheme 6.9. Of all the high nuclearity clusters of cobalt, this is the most simple to prepare, and it has been used extensively in reactivity studies.

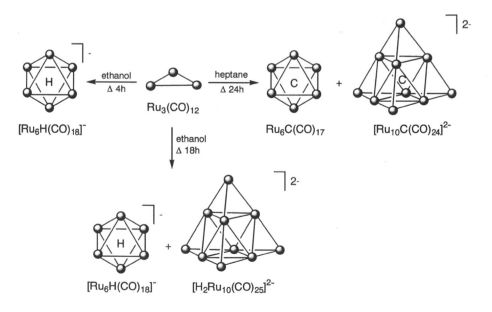

SCHEME 6.8 SOME PRODUCTS OBTAINED FROM THE THERMOLYSIS OF $Ru_3(CO)_{12}$.

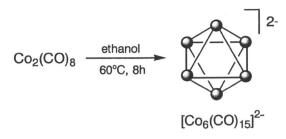

$Co_2(CO)_8 \xrightarrow[\text{60°C, 8h}]{\text{ethanol}}$

$[Co_6(CO)_{15}]^{2-}$

SCHEME 6.9 THE SYNTHESIS OF $[Co_6(CO)_{15}]^{2-}$.

With iron clusters, thermolysis alone is not sufficient to bring about the formation of higher nuclearity species and reducing agents must be added to the reaction mixture in order to promote cluster growth. An example involves heating iron pentacarbonyl with $[Mn(CO)_5]^-$ in refluxing diglyme to afford the octahedral cluster $[Fe_6C(CO)_{16}]^{2-}$ in 65% yield, as illustrated in Scheme 6.10. Acidification of this cluster causes it to undergo fragmentation to give the square-pyramidal cluster $Fe_5C(CO)_{15}$. A related cluster with a similar metal atom topology, but containing a nitride atom in place of the carbide is $[Fe_5N(CO)_{14}]^-$. This cluster is prepared from the reaction of $Na_2[Fe_2(CO)_8]$, $Fe(CO)_5$ and $NOBF_4$.

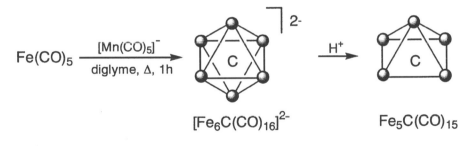

$Fe(CO)_5 \xrightarrow[\text{diglyme, } \Delta, \text{ 1h}]{[Mn(CO)_5]^-}$    $[Fe_6C(CO)_{16}]^{2-}$  $\xrightarrow{H^+}$  $Fe_5C(CO)_{15}$

SCHEME 6.10 THE SYNTHESIS OF $[Fe_6C(CO)_{16}]^{2-}$ AND SUBSEQUENT CONVERSION TO $Fe_5C(CO)_{15}$.

In contrast to iron and ruthenium clusters the most effective method used to prepare high nuclearity clusters of osmium is pyrolysis. The method simply involves heating a low nuclearity cluster in a sealed, evacuated tube. High temperatures are generally required and reactions can take between a few hours and several days. An example of a vacuum pyrolysis is that of $Os_3(CO)_{12}$, which results in the formation of a number of clusters including $Os_5(CO)_{16}$, $Os_6(CO)_{18}$, $Os_7(CO)_{21}$, $Os_8(CO)_{23}$, $Os_5C(CO)_{15}$, $Os_5C(CO)_{16}$, $Os_8C(CO)_{22}$ and $[Os_{10}C(CO)_{24}]^{2-}$. The precise range and distribution of products depends upon the temperature and time of the pyrolysis.

The wide number of clusters prepared in these types of reactions is very impressive. However, the ability to prepare a single cluster in high yield is also desirable, as it then becomes possible to study the chemistry of the cluster in detail. Many synthetic methods have been optimised with this goal in mind and certain pyrolysis reactions are highly selective. For example, $Os_6(CO)_{18}$ can be prepared by vacuum pyrolysis in yields exceeding 80%. It is, however, very difficult to envisage why products such as $Os_6(CO)_{18}$ are obtained in the pyrolysis reaction and a more selective and predictable technique for the preparation of high nuclearity clusters is desirable. A more selective reaction is the redox condensation reaction and this is described in Section 6.2.4.

### Pyrolysis products of $Os_3(CO)_{10}(NCMe)_2$

It is often preferable to activate the cluster prior to vacuum pyrolysis, *i.e.* modify the cluster so that it is more reactive. With $Os_3(CO)_{12}$, this can be achieved by substitution of two carbonyl ligands by two comparatively weakly coordinated ligands, such as acetonitrile, to afford $Os_3(CO)_{10}(NCMe)_2$. These relatively labile ligands make $Os_3(CO)_{10}(NCMe)_2$ more reactive than homoleptic species. Pyrolysis of the *bis*-substituted acetonitrile cluster $Os_3(CO)_{10}(NCMe)_2$ affords a large range of products depending upon the conditions employed. Vacuum pyrolysis of $Os_3(CO)_{10}(NCMe)_2$ has been carried out at temperatures between 100 and 300°C over time periods ranging from hours to days. Thirteen different clusters have been extracted and characterised from these reactions so far, and it is entirely likely that even more new compounds could be identified if the reaction conditions are varied further. Some of the high nuclearity clusters isolated in this way are shown.

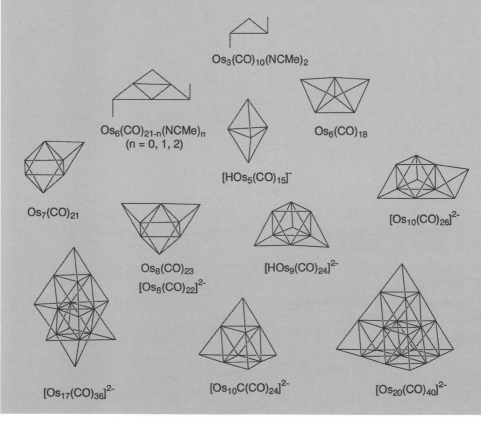

$Os_3(CO)_{10}(NCMe)_2$

$Os_6(CO)_{21-n}(NCMe)_n$
$(n = 0, 1, 2)$

$Os_6(CO)_{18}$

$[HOs_5(CO)_{15}]^-$

$Os_7(CO)_{21}$

$[Os_{10}(CO)_{26}]^{2-}$

$Os_8(CO)_{23}$
$[Os_8(CO)_{22}]^{2-}$

$[HOs_9(CO)_{24}]^{2-}$

$[Os_{17}(CO)_{36}]^{2-}$

$[Os_{10}C(CO)_{24}]^{2-}$

$[Os_{20}(CO)_{40}]^{2-}$

### 6.2.3 Reduction and oxidation

A large number of nickel clusters have been prepared from the reduction of nickel tetracarbonyl by using a selection of reducing agents under various conditions. For example, $[Ni_6(CO)_{12}]^{2-}$ may be prepared from $Ni(CO)_4$ by the action of various reducing agents including lithium, sodium, potassium, magnesium, NaOH, NaBH$_4$ and KOH. The hexanickel cluster is obtained in particularly high yield when nickel tetracarbonyl is added in small portions to a solution of sodium hydroxide in DMSO over several hours or by heating $Ni(CO)_4$ with sodium borohydride for two hours. The tetraanion $[Ni_{12}(CO)_{21}]^{4-}$ is isolated as a by-product with $[Ni_6(CO)_{12}]^{2-}$ (the major product) when nickel tetracarbonyl is reduced with KOH in methanol over a prolonged period. These reactions, and other examples, are illustrated in Scheme 6.11.

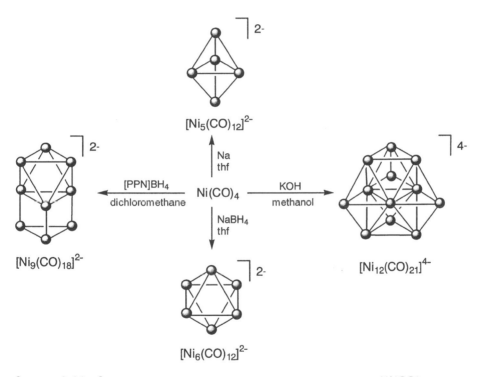

SCHEME 6.11  SOME PRODUCTS OBTAINED FROM THE REDUCTION OF $Ni(CO)_4$.

Reduction is the most effective technique for the preparation of high nuclearity rhodium clusters. The preparation of $Rh_4(CO)_{12}$ and $Rh_6(CO)_{16}$ from $RhCl_3$ was described earlier in this chapter. These compounds have been widely used as precursors to a large number of higher nuclearity clusters. For example, reduction of $Rh_4(CO)_{12}$ with sodium hydroxide in propanol affords $[Rh_{17}(CO)_{30}]^{3-}$ in 20% yield, whereas the reduction of $Rh_6(CO)_{16}$ with potassium hydroxide in methanol results in only a slight increase in nuclearity with the formation of $[Rh_7(CO)_{16}]^{3-}$. These reactions are illustrated in Scheme 6.12.

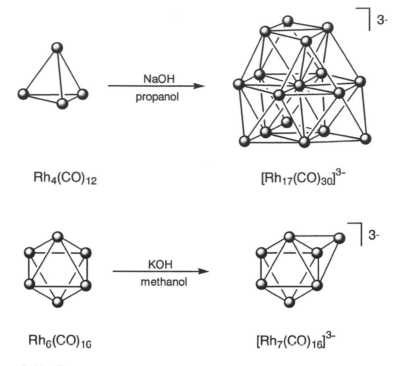

Rh$_4$(CO)$_{12}$       NaOH, propanol       [Rh$_{17}$(CO)$_{30}$]$^{3-}$

Rh$_6$(CO)$_{16}$       KOH, methanol       [Rh$_7$(CO)$_{16}$]$^{3-}$

SCHEME 6.12   THE SYNTHESIS OF RHODIUM CLUSTERS BY REDUCTION.

The most frequently encountered high nuclearity iridium cluster is [Ir$_6$(CO)$_{15}$]$^{2-}$. It has been widely used as a starting material in many reactions. It is prepared from Ir$_4$(CO)$_{12}$ by reduction with sodium metal and depending upon the precise conditions employed the yield varies from 35–65%. The reduction of Ir$_4$(CO)$_{12}$ with an excess of potassium hydroxide in methanol affords the bicapped octahedral cluster [Ir$_8$(CO)$_{22}$]$^{2-}$ in about 60% yield. These reactions are depicted in Scheme 6.13.

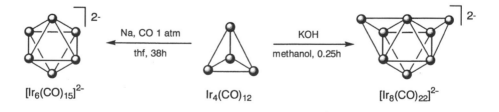

[Ir$_6$(CO)$_{15}$]$^{2-}$    Na, CO 1 atm, thf, 38h    Ir$_4$(CO)$_{12}$    KOH, methanol, 0.25h    [Ir$_8$(CO)$_{22}$]$^{2-}$

SCHEME 6.13   THE SYNTHESIS OF HIGH NUCLEARITY IRIDIUM CLUSTERS BY REDUCTION OF Ir$_4$(CO)$_{12}$.

There are also a few examples of oxidation reactions that result in cluster growth, mainly for anionic rhodium and nickel clusters. Examples include the reaction of $[Rh_6C(CO)_{15}]^{2-}$ with ferric ammonium alum, which gives $Rh_8C(CO)_{19}$ or $[Rh_{15}C_2(CO)_{28}]^-$, depending on whether the reaction is carried out under an atmosphere of carbon monoxide or nitrogen (see Scheme 6.14). Oxidation of $[Rh_6C(CO)_{15}]^{2-}$ with sulphuric acid results in the formation of $[Rh_{12}C_2(CO)_{24}]^{2-}$ in 90% yield.

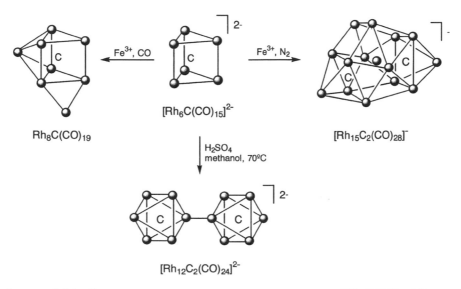

SCHEME 6.14 OXIDATIVE CLUSTER BUILD-UP REACTIONS FROM $[Rh_6C(CO)_{15}]^{2-}$.

### 6.2.4 Redox condensation reactions

The redox condensation reaction is regarded as the best method for targeting particular cluster compounds. It is a type of metathesis reaction and is especially suited to the preparation of mixed metal (heteronuclear) clusters (see Chapter 9). In general, all that is required is careful selection of the starting materials. For example, the reaction between the dianionic cluster $[Rh_6C(CO)_{15}]^{2-}$ and the dicationic complex $[Rh(CO)_2(NCMe)_2]^{2+}$ affords $[Rh_{14}C_2(CO)_{33}]^{2-}$ in good yield. This cluster consists of two monocapped trigonal prisms linked through two capping rhodium atoms (see Scheme 6.15).

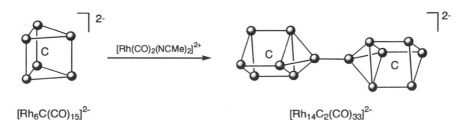

SCHEME 6.15 THE SYNTHESIS OF $[Rh_{14}C_2(CO)_{33}]^{2-}$ USING A REDOX CONDENSATION REACTION.

Not all redox condensation reactions are as easily rationalised as the one described above. A more unpredictable example involves the reaction between the neutral species $[\mu_4\text{-Si}\{Co_2(CO)_7\}_2]$ and an excess of the anion $[Co_4(CO)_4]^-$ to afford $[Co_9Si(CO)_{21}]^{2-}$ (see Scheme 6.16). Nevertheless, the reaction product is extremely interesting and an alternative route to $[Co_9Si(CO)_{21}]^{2-}$ is not obvious. The nonacobalt cluster has a monocapped square antiprismatic core with the interstitial silicon atom residing in the centre of the cavity.

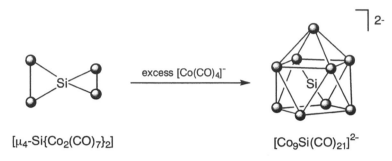

$[\mu_4\text{-Si}\{Co_2(CO)_7\}_2]$        excess $[Co(CO)_4]^-$        $[Co_9Si(CO)_{21}]^{2-}$

SCHEME 6.16  THE SYNTHESIS OF $[Co_9Si(CO)_{21}]^{2-}$ BY REDOX CONDENSATION.

### 6.2.5 Cluster build-up reactions from preformed high nuclearity clusters

There are a large number of compounds that can be prepared from the condensation of high nuclearity clusters, and examples have already been described, including the oxidation of the octahedral cluster $[Rh_6C(CO)_{15}]^{2-}$ with sulphuric acid to form $[Rh_{12}C_2(CO)_{24}]^{2-}$ (Scheme 6.14) and the redox condensation reaction shown in Scheme 6.15. In general these reactions tend to involve the thermolysis of a high nuclearity cluster to form an even larger cluster. Further examples include the thermolysis of $[Co_6C(CO)_{15}]^{2-}$ in diglyme to afford $[Co_{13}C_2(CO)_{24}]^{4-}$, the thermolysis of $[Ru_6C(CO)_{16}]^{2-}$ in tetraglyme to afford $[Ru_{10}C_2(CO)_{24}]^{2-}$ and the thermolysis of $[Pt_6(CO)_{12}]^{2-}$ in acetonitrile that gives $[Pt_{19}(CO)_{22}]^{2-}$. These reactions are illustrated in Scheme 6.17.

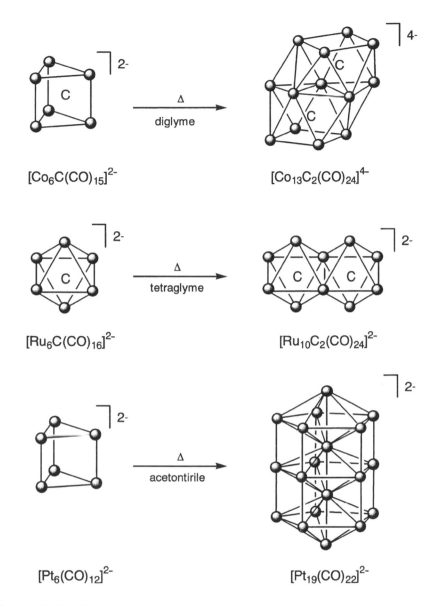

$[Co_6C(CO)_{15}]^{2-}$         $[Co_{13}C_2(CO)_{24}]^{4-}$

$[Ru_6C(CO)_{16}]^{2-}$         $[Ru_{10}C_2(CO)_{24}]^{2-}$

$[Pt_6(CO)_{12}]^{2-}$         $[Pt_{19}(CO)_{22}]^{2-}$

SCHEME 6.17 CLUSTER GROWTH REACTIONS STARTING WITH HIGH NUCLEARITY CLUSTER PRECURSORS.

CHAPTER 7

# REACTIVITY

This chapter outlines the main types of reactions that transition metal carbonyl clusters undergo. Attention is focused on the differences in reactivity between clusters and mononuclear carbonyl complexes. These differences are traced to the specific bonding requirements of clusters. Mechanistic details illustrating the importance of metal-metal bond breaking/reformation in reactions are described. The last two sections deal with carbide formation and cluster decomposition reactions that are unique to polynuclear compounds.

## 7.1 REDUCTION

Transition metal carbonyl clusters are generally easy to reduce as the additional charge can be 'spread out' over several metals as well as being further delocalised onto the carbonyl ligands due to increased back bonding (see Chapter 4). As the number of metals in the cluster increases very high negative charges can be attained. However, although the overall charge may be low, say −10, when divided by the number of metals actually present, the individual oxidation state of each metal ion is usually between 0 and −1.

Reduction is a very important reaction in organometallic chemistry. The classical route used to reduce a metal carbonyl compound involves reaction of the neutral compound with a strong base. For example, the reduction of $Fe(CO)_5$, $Fe_2(CO)_9$ and $Fe_3(CO)_{12}$ with $OH^-$ results in the addition of two electrons to the compound, accompanied by loss of a CO ligand. The nucleophile ($OH^-$) attacks the electropositive carbon atom of a carbonyl ligand and the initial addition product decomposes by $\beta$-elimination to give the carbonylmetallate shown in Scheme 7.1. It should be noted that in the reactions of $Fe(CO)_5$, $Fe_2(CO)_9$ and $Fe_3(CO)_{12}$ with $OH^-$, the nuclearity of the complexes on conversion to their corresponding dianions $[Fe(CO)_4]^{2-}$, $[Fe_2(CO)_8]^{2-}$ and $[Fe_3(CO)_{11}]^{2-}$, is not affected.

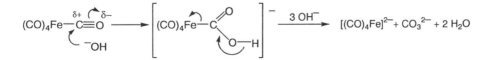

SCHEME 7.1 THE MECHANISM FOR THE REDUCTION OF METAL CARBONYLS USING $OH^-$.

The reduction of other clusters tends to be more complicated and may involve several different pathways. The actual route depends upon both the cluster being reduced and the reducing agent employed. The different pathways comprise:

i.   Loss of CO.
ii.  Breaking of a metal-metal bond.
iii. Polyhedral rearrangement.
iv.  Changes to the nuclearity of the cluster.
v.   No structural change.

Examples to illustrate points i to v will be given in turn.

Reduction of the square pyramidal cluster $Ru_5C(CO)_{15}$ with an excess of KOH in methanol results in loss of CO affording $[Ru_5C(CO)_{14}]^{2-}$ as shown in Scheme 7.2. The mechanism of this reaction proceeds in a similar manner to iron carbonyl illustrated in Scheme 7.1. The dianionic product $[Ru_5C(CO)_{14}]^{2-}$ is used in redox condensation reactions as the square face is ideally suited to being capped by a number of mononuclear cationic fragments (see Chapter 9).

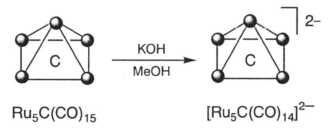

$$Ru_5C(CO)_{15} \qquad [Ru_5C(CO)_{14}]^{2-}$$

SCHEME 7.2  THE REDUCTION OF $Ru_5C(CO)_{15}$ RESULTING IN LOSS OF CO.

Reduction that results in breaking a metal-metal bond is best illustrated with dinuclear complexes. For example, the reduction of $Co_2(CO)_8$ or $Mn_2(CO)_{10}$ with two equivalents of sodium affords $[Co(CO)_4]^-$ and $[Mn(CO)_5]^-$, respectively. The resulting mononuclear anions are useful reagents in synthesis, including the preparation of heteronuclear (mixed-metal) clusters (see Chapter 9).

With higher nuclearity clusters the skeletal bonding is described by a delocalised bonding model, rather than in terms of two centre/two electron metal-metal bonds (see Chapter 2). The more flexible bonding of the former allows complicated polyhedral transformations to take place upon reduction. A prominent example of such a process involves the reduction of $Os_6(CO)_{18}$ to $[Os_6(CO)_{18}]^{2-}$, illustrated in Scheme 7.3. The neutral hexaosmium cluster has a bicapped tetrahedral geometry whereas the dianionic cluster has an octahedral osmium core. The reduction of $Os_6(CO)_{18}$ to $[Os_6(CO)_{18}]^{2-}$ can be achieved chemically or electrochemically. It is a *pseudo*-first order process (established by spectroelectrochemistry) and the structural rearrangement takes place during the first electron-transfer step. The reverse reaction is less straightforward, but it would appear that the structural rearrangement takes place after the oxidation process is complete. This polyhedral transformation can be rationalised in terms of the total electron counts for the two clusters. The neutral cluster $Os_6(CO)_{18}$ has a total electron count of 84, which corresponds to a bicapped tetrahedral geometry whereas the dianion has two more electrons, i.e. 86, which corresponds to the electron count of an octahedron.

In certain clusters, reduction can take place without the loss of carbonyl ligands

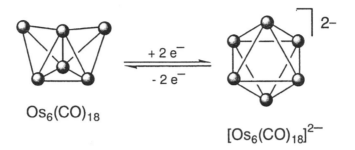

$$Os_6(CO)_{18} \quad \xrightleftharpoons[-2\,e^-]{+2\,e^-} \quad [Os_6(CO)_{18}]^{2-}$$

SCHEME 7.3   THE REVERSIBLE REDUCTION OF $Os_6(CO)_{18}$ INVOLVING A POLYHEDRAL TRANSFORMATION.

and without changes to the metal core. In fact, certain clusters are referred to as 'electron reservoirs' because of their ability to undergo reduction to very high negative charges, −10 and beyond. In general, the larger the cluster the greater the tolerance it has to be reduced to a high negative charge, because the charge can be distributed (or delocalised) over a greater number of metals and their corresponding carbonyls. A series of large nickel clusters are particularly good electron reservoirs. The clusters $[Ni_{32}C_6(CO)_{36}]^{n-}$ and $[Ni_{38}C_6(CO)_{42}]^{n-}$ (see Figure 7.1) have a very wide range of oxidation states. In the former cluster, the charge ranges from −5 to −10, and in

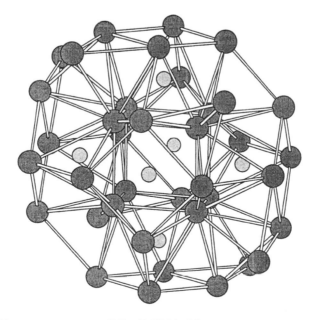

FIGURE 7.1   THE STRUCTURE OF $[Ni_{38}C_6(CO)_{42}]^{5-}_n$.

$[Ni_{38}C_6(CO)_{42}]^{n-}$, the charge varies from $-5$ to $-9$. The stability of the metal core in these clusters is due to the presence of the interstitial carbide atoms (see Chapter 4) and the high coordination number of the nickel atoms. Calculations have shown that $[Ni_{32}C_6(CO)_{36}]^{n-}$ and $[Ni_{38}C_6(CO)_{42}]^{n-}$ bear some qualitative resemblance to nanoscale metal particles in that they are close to having actual conduction bands.

With increasing nuclearity the separation between the high bonding and lowest antibonding skeletal molecular orbitals decreases. High nuclearity clusters have many closely spaced energy levels and approach the d-d bandwidth (band gap) of a bulk metal.

### 7.2   PROTONATION

Transition metal carbonyl compounds containing hydride ligands have been intensively studied as they represent intermediates in many metal catalysed reactions involving CO and $H_2$. The origin of the hydride ligands in many clusters is from hydrogen atoms derived from the solvent molecules in which the cluster is prepared. Alternatively, the cluster precursor might have contained hydrides and these are carried over into the products. Examples of these types of reactions were given in Chapter 6. However, hydrides can also be introduced using a rational synthetic route, the most common of which involves acidification of an anionic cluster.

Metal carbonyl hydrides can be viewed as the conjugate acids from which the anions are derived by proton displacement, despite the term hydride suggesting $H^-$ character. In fact, some hydride complexes behave like acids although poor solubility in water often makes it difficult to quantify their acidic properties. In general, protonation of mononuclear carbonyl anions with acids is straightforward. For example, $[Co(CO)_4]^-$ reacts with acid to afford $HCo(CO)_4$. However, with clusters more possibilities exist and some illustrative examples are described below.

The decanuclear cluster $[Ru_{10}C(CO)_{24}]^{2-}$ may be protonated with HCl to give $[HRu_{10}C(CO)_{24}]^-$, whereas other acids including sulphuric and trifluoroacetic acid cause the cluster to decompose. The precise location of the hydride ligand in this cluster is not known. Similarly, the hexacobalt cluster $[Co_6(CO)_{15}]^{2-}$ reacts with HCl to form $[Co_6H(CO)_{15}]^-$. The hydride in this cluster occupies the interstitial site as shown in Scheme 7.4.

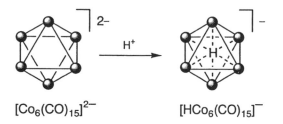

$[Co_6(CO)_{15}]^{2-}$          $[HCo_6(CO)_{15}]^-$

SCHEME 7.4   PROTONATION OF $[Co_6(CO)_{15}]^{2-}$.

Acidification of cluster anions does not always result in the hydrogenation of the cluster. For example, treatment of $[Fe_6C(CO)_{16}]^{2-}$ with sulphuric acid affords $Fe_5C(CO)_{15}$ as shown in Scheme 7.5. This reaction may be described as oxidative decapping, or oxidative decomposition, and some more examples of decomposition reactions are given in Section 7.7.

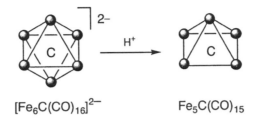

$$[Fe_6C(CO)_{16}]^{2-} \qquad Fe_5C(CO)_{15}$$

SCHEME 7.5   OXIDATIVE DECAPPING OF $[Fe_6C(CO)_{16}]^{2-}$.

Molecular hydrogen is cleaved in certain reactions with clusters to form hydride compounds. Bubbling $H_2$ gas though a refluxing octane solution of $Ru_3(CO)_{12}$ for several hours affords the tetranuclear cluster $H_4Ru_4(CO)_{12}$ (see Scheme 7.6). Under almost identical reaction conditions $Os_3(CO)_{12}$ reacts to form the electronically unsaturated triosmium cluster $H_2Os_3(CO)_{10}$. The tetraosmium analogue of the ruthenium cluster, i.e. $H_4Os_4(CO)_{12}$, can be prepared, but only under more forcing reaction conditions as shown in the Scheme.

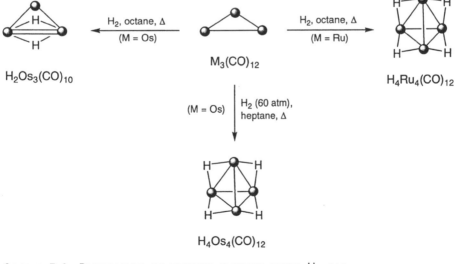

$$H_2Os_3(CO)_{10} \qquad\qquad M_3(CO)_{12} \qquad\qquad H_4Ru_4(CO)_{12}$$

$$H_4Os_4(CO)_{12}$$

SCHEME 7.6   PREPARATION OF HYDRIDE CLUSTERS USING $H_2$ GAS.

A far less selective method used to prepare cluster-hydride complexes involves carrying out the synthesis of the cluster in the presence of a hydride source. An example is the vacuum pyrolysis of $Os_3(CO)_{12}$ to which small amount of water is deliberately added. A number of hydride clusters are obtained from this reaction including $H_2Os_5(CO)_{15}$, $H_2Os_5(CO)_{16}$, $H_2Os_6(CO)_{18}$, $H_2Os_7C(CO)_{19}$, $H_2Os_{10}C(CO)_{24}$ and $H_2Os_{11}C(CO)_{27}$.

## 7.3 SUBSTITUTION

Substitution reactions are among the most widely studied reactions in transition metal cluster chemistry. There is a huge range of potential ligands that can be used and examples involving unsaturated organic molecules will be discussed in Chapter 8. Here, most attention will be paid to phosphine nucleophiles, because there is considerable mechanistic data available for phosphine substitution in metal carbonyls. As such, the underlying mechanisms involved in phosphine substitution reactions in clusters are well understood.

Substitution of CO by phosphine nucleophiles in $Ru_3(CO)_{12}$ involves an associative reaction mechanism. In fact, $Ru_3(CO)_{12}$ is several orders of magnitude more susceptible to nucleophilic attack than most mononuclear homoleptic complexes. The reason is believed to be the ability of the cluster to undergo Ru-Ru bond breaking, concurrently with formation of the ruthenium ligand bond, resulting in the intermediate shown in Scheme 7.7.

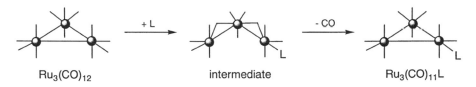

$Ru_3(CO)_{12}$     intermediate     $Ru_3(CO)_{11}L$

SCHEME 7.7 THE ASSOCIATIVE MECHANISM FOR LIGAND SUBSTITUTION IN $Ru_3(CO)_{12}$.

The intermediate maintains an 18-electron count at each metal centre and subsequent loss of a CO ligand is accompanied by reformation of the Ru-Ru bond and isolation of the substitution product. An alternative mechanism has been observed to occur with very strong nucleophiles, involving the complete fragmentation of $Ru_3(CO)_{12}$ into mononuclear species. Subsequent association of the new ligand and reformation of the cluster with concurrent expulsion of a carbonyl gives the appropriate derivative. In fact, in certain substitution reactions cluster growth is also observed, suggesting that complete fragmentation also takes place in this instance.

The mechanism by which substitution reactions take place in $Ru_3(CO)_{12}$ is governed by the basicity of the ligand. With poorer nucleophiles substitution occurs via an associative mechanism and with stronger nucleophiles substitution occurs via a dissociative mechanism. In contrast, the mechanism of substitution in $Ru_5C(CO)_{15}$ for a range of phosphine ligands is determined by the cone angle ($\theta$) of the ligand. Ligands with cone angles of 133° or less react rapidly to form $Ru_5C(CO)_{14}L$ in an associative two step process, as shown in Scheme 7.8. Spectroscopic evidence demonstrated that in the first step the association product $Ru_5C(CO)_{15}L$ was formed. A basal-to-apex Ru-Ru bond in the square pyramid breaks to give a so-called 'bridged butterfly' structure. This type of metal core transformation is well known for $Ru_5C(CO)_{15}$, and further examples are described in Section 7.4. In the second step, loss of CO is accompanied by reformation of the Ru-Ru bond to produce the square pyramidal cluster with a different ligand attached to the basal ruthenium atom. This reaction is very similar to that described for $Ru_3(CO)_{12}$. However, larger ligands with cone angles exceeding $136°$ react much more slowly than the smaller ligands. There is no evidence for the formation of a bridged butterfly intermediate and the reaction takes place in a single step. The actual mechanism for this single step process is believed to involve breaking a basal-basal Ru-Ru bond to form a different adduct, which loses CO so rapidly that it cannot be detected.

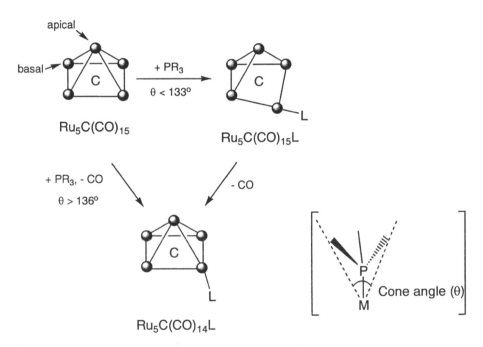

SCHEME 7.8   LIGAND SUBSTITUTION MECHANISMS IN $Ru_5C(CO)_{15}$.

Carbonyls form a relatively strong bond to transition metals compounds compared to ligands such as acetonitrile, which is a poor $\pi$-acceptor. As such, exchanging a CO ligand for an acetonitrile ligand is a common route used to make a cluster more reactive. For example, a carbonyl ligand in $Os_3(CO)_{12}$ can be removed as $CO_2$ by reaction with trimethylamine-$N$-oxide ($Me_3NO$) (see Scheme 7.9). $Me_3NO$ is a nucleophile and attacks the electropositive carbon atom of the carbonyl to produce $Me_3N$ and $CO_2$. The $Me_3N$ weakly coordinates to the vacant coordination site produced on the cluster and is readily displaced by another ligand. If this reaction is carried out in the presence of acetonitrile the substituted product $Os_3(CO)_{11}(NCMe)$ is produced. This type of cluster is often referred to as 'activated', in the sense that the weakly coordinated MeCN ligand is readily displaced in a simple dissociative process by a wide range of other ligands under ambient conditions.

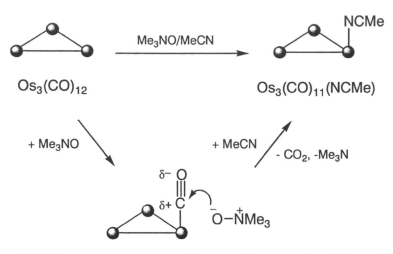

SCHEME 7.9 THE PREPARATION OF THE 'ACTIVATED' CLUSTER $Os_3(CO)_{11}(NCMe)$.

Another cluster that contains an acetonitrile ligand is the raft cluster $Os_6(CO)_{20}(NCMe)$, derived from $Os_6(CO)_{21}$ (see Figure 7.2). The substitution of the MeCN ligand in this raft does not follow a simple dissociative pathway. It has been found that these raft clusters contain a low-lying empty molecular orbital. This empty orbital readily accepts an electron pair from a suitable 2-electron donor ligand, thereby forming $Os_6(CO)_{20}(NCMe)L$. However, this adduct loses MeCN in a dissociative step very rapidly so that the intermediate adduct is not detected. This reaction represents a further example where substitution involves initial association of the new ligand. In this case, however, no metal-metal bond breaking is observed, and while the intermediate is too short lived to be detected, the rate of reaction is considerably faster than in corresponding complexes that do not contain an empty, low-lying molecular orbital.

## 7.4 ASSOCIATION AND DISSOCIATION

Association reactions are commonplace in mononuclear organometallic chemistry when, for example, a coordinatively unsaturated 16-electron complex reacts with a 2-electron donor ligand to afford a saturated 18-electron species. In clusters, ligand association is also common and occurs when the cluster is electronically unsaturated

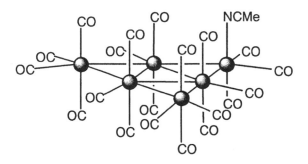

FIGURE 7.2 THE STRUCTURE OF $Os_6(CO)_{20}(NCMe)$.

or where cleavage of a metal-metal bond or polyhedral rearrangement is relatively facile. The reverse process, i.e. dissociation, is also relatively common, especially in the processes involving changes to the cluster core.

A simple example of association involves the reaction of $H_2Os_3(CO)_{10}$ with phosphines, $PR_3$, to give the adduct $H_2Os_3(CO)_{10}(PR_3)$ (see Scheme 7.10). This reaction is easily rationalised since $H_2Os_3(CO)_{10}$ is electronically unsaturated. It has 46 valence electrons, two fewer than that expected for an electronically saturated triangular cluster and the edge bridged by the two hydrides can be viewed as containing an Os=Os double bond. The product has 48 valence electrons and as such subsequent dissociation of the phosphine (or any of the carbonyl ligands) does not readily take place.

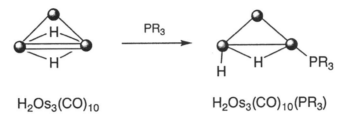

$$H_2Os_3(CO)_{10} \qquad H_2Os_3(CO)_{10}(PR_3)$$

SCHEME 7.10   ASSOCIATION OF A PHOSPHINE ACROSS THE 'DOUBLE BOND' IN $H_2Os_3(CO)_{10}$.

The substitution chemistry of $Ru_5C(CO)_{15}$ with different phosphine ligands was described in Section 7.3. This cluster also undergoes association and dissociation reactions with appropriate ligands such as CO and acetonitrile (see Scheme 7.11). Association of CO takes place with concomitant breaking of a basal-to-apex Ru-Ru bond giving a bridged butterfly metal core. Dissociation of the ligand results in the reformation of the original square pyramidal cluster core.

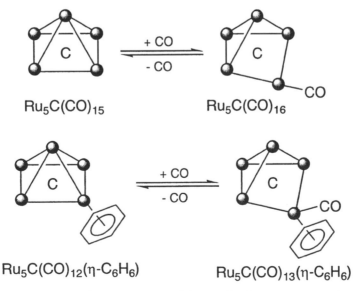

$$Ru_5C(CO)_{15} \qquad Ru_5C(CO)_{16}$$

$$Ru_5C(CO)_{12}(\eta\text{-}C_6H_6) \qquad Ru_5C(CO)_{13}(\eta\text{-}C_6H_6)$$

SCHEME 7.11   ASSOCIATION/DISSOCIATION OF CO WITH Ru-Ru BOND SCISSION/FORMATION IN $Ru_5C(CO)_{15}$ AND $Ru_5C(CO)_{12}(\eta\text{-}C_6H_6)$.

The benzene derivative shown in Scheme 7.11, i.e. $Ru_5C(CO)_{12}(\eta\text{-}C_6H_6)$, also undergoes association reactions with ligands such as CO. The rate at which this happens is somewhat faster than that of the precursor, $Ru_5C(CO)_{15}$. The reason for the increased rate is that the benzene ligand, which is attached to a basal ruthenium atom, increases the electrophilic character at the ruthenium centre, because benzene is a poorer $\pi$-ligand than the three CO ligands it has replaced. As such, the more electropositive ruthenium atom is more susceptible to nucleophilic attack.

The butterfly cluster $Pd_4(CO)_5(PBu^n_3)_4$ undergoes an association reaction with CO under ambient conditions to form the tetrahedral $Pd_4(CO)_6(PBu^n_3)_4$ (see Scheme 7.12). What is surprising about this reaction is that a butterfly cluster converts to a tetrahedral cluster with the association of a 2-electron donor ligand. Tetrahedral clusters usually have two fewer valence electrons than butterfly clusters (i.e. 60 and 62 cluster valence electrons, respectively, see Chapter 2). As such, this transformation would be more typical of ligand loss rather than association. However, this anomaly is readily attributed to the ability of palladium to form stable compounds with 14, 16 and 18 valence electrons. As such, $Pd_4(CO)_5(PBu^n_3)_4$ and $Pd_4(CO)_6(PBu^n_3)_4$ do not conform to the usual total electron counts for compounds which obey the EAN rule, but have 54 and 56 electrons respectively.

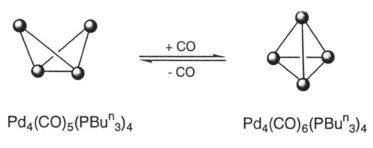

$$Pd_4(CO)_5(PBu^n_3)_4 \qquad\qquad Pd_4(CO)_6(PBu^n_3)_4$$

SCHEME 7.12  THE TRANSFORMATION OF THE METAL CORE ON ASSOCIATION/DISSOCIATION OF CO.

The heteronuclear cluster $Ru_6Pt_3(CO)_{14}(\mu_3\text{-}C_2Ph_2)_3$ reacts with four equivalents of carbon monoxide to give $Ru_6Pt_3(CO)_{18}(\mu_3\text{-}C_2Ph_2)_3$. The association of the four CO ligands is accompanied by a considerable polyhedral rearrangement, which is illustrated in Scheme 7.13. The reaction can be reversed by thermally induced dissociation of the four carbonyl ligands.

## 7.5  OXIDATIVE ADDITION AND REDUCTIVE ELIMINATION

In the previous section some association and dissociation reactions of clusters were described. If, however, the oxidation state of the metals change during association and dissociation, the reactions are termed oxidative addition and reductive elimination, respectively. These reactions are very important in many catalytic processes and a few examples of well characterised reactions will be described here. Further examples of oxidative addition and reductive elimination reactions involved in actual catalytic processes are described in Chapter 10.

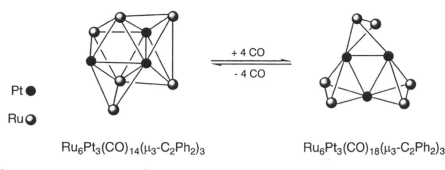

Pt ●

Ru ○

$Ru_6Pt_3(CO)_{14}(\mu_3\text{-}C_2Ph_2)_3$       $Ru_6Pt_3(CO)_{18}(\mu_3\text{-}C_2Ph_2)_3$

SCHEME 7.13   ASSOCIATION/DISSOCIATION OF FOUR CO LIGANDS INVOLVING A POLYHEDRAL REARRANGEMENT.

One of the most commonly encountered types of oxidative addition reaction in clusters involves the oxidative addition of CO or some other 2-electron donor ligand to a dianionic cluster. For example, $[Ir_6(CO)_{15}]^{2-}$ undergoes oxidative addition with MeCOOH in the presence of CO to afford $Ir_6(CO)_{16}$. The average oxidation state of each Os-atom in $[Ir_6(CO)_{15}]^{2-}$ is $-0.33$ and in $Ir_6(CO)_{16}$, it is zero. Reduction of $Ir_6(CO)_{16}$ to $[Ir_6(CO)_{15}]^{2-}$, or more strictly, reductive elimination, can be achieved by reaction with a number of different bases.

The redox chemistry of $Ru_6C(CO)_{17}$ and $[Ru_6C(CO)_{16}]^{2-}$ has been probed using spectroelectrochemistry. This technique involves investigating the electrochemistry of a compound while simultaneously monitoring the IR spectrum. When $Ru_6C(CO)_{17}$ is reduced under an inert atmosphere $[Ru_6C(CO)_{16}]^{2-}$ and CO are formed, as expected. This is clearly reductive elimination as reduction is accompanied by the expulsion of a CO ligand. However, the changes in the IR spectrum accompanying the reduction of $Ru_6C(CO)_{17}$ in the presence of oxygen show that reduction is accompanied by the formation of $[Ru_6C(CO)_{16}]^{2-}$ and $CO_2$. Although the reason for the formation of $CO_2$ is not fully understood, the potential that this system has as a catalyst for the conversion of waste CO into $CO_2$ is apparent as $[Ru_6C(CO)_{16}]^{2-}$ readily undergoes oxidative addition with CO to regenerate $Ru_6C(CO)_{17}$ as shown in Scheme 7.14.

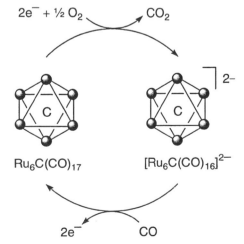

$2e^- + \tfrac{1}{2} O_2$     $CO_2$

$Ru_6C(CO)_{17}$       $[Ru_6C(CO)_{16}]^{2-}$

$2e^-$     CO

SCHEME 7.14   THE REACTION CYCLE FOR THE CONVERSION OF CO TO $CO_2$ INVOLVING REDUCTIVE ELIMINATION AND OXIDATIVE ADDITION.

The tetrahedral cluster $[H_2Os_4(CO)_{12}I]^-$ readily undergoes oxidative addition on protonation to afford the butterfly cluster $H_3Os_4(CO)_{12}(\mu\text{-}I)$, according to Scheme 7.15. In this reaction the bonding mode of the iodine changes from terminal to bridging, spanning the gap between the wing-tip atoms of the butterfly. The change in bonding mode is accompanied by an increase in the number of electrons the iodine atom contributes towards cluster bonding, from 1 to 3. The change is driven by the electronic demands of the butterfly cluster, which requires a total electron count of 62. The average oxidation state of each osmium atom in $[H_2Os_4(CO)_{12}I]^-$ is zero and changes to +0.5 in $H_3Os_4(CO)_{12}(\mu\text{-}I)$ (remember H is formally +1 and I is formally -1). Reductive elimination of $H_3Os_4(CO)_{12}(\mu\text{-}I)$ is achieved by deprotonation with a suitable base, which regenerates the tetrahedral 60 electron anion $[H_2Os_4(CO)_{12}I]^-$.

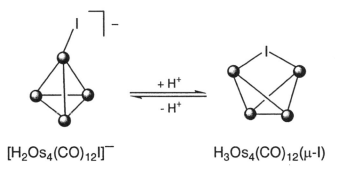

$$[H_2Os_4(CO)_{12}I]^- \qquad\qquad H_3Os_4(CO)_{12}(\mu\text{-}I)$$

SCHEME 7.15   OXIDATIVE ADDITION/REDUCTIVE ELIMINATION OF A TETRAOSMIUM CLUSTER.

## 7.6   CARBIDE FORMATION

The study of carbonyl clusters containing interstitial main group atoms is an important area of cluster chemistry. The introduction of a p-block element into a cavity in the cluster skeleton has three purposes:

i.    To polarise bonding within the cluster core leading to different reactivity compared to clusters that do not contain an interstitial atom.
ii.   To act as an inert centre stabilising the cluster core towards the conditions required for stoichiometric and catalytic reactions.
iii.  To provide extra electrons for bonding.

Interstitial atoms such as carbon are often derived from CO. The ability of clusters to break a C≡O bond under relatively mild conditions is quite remarkable. Understanding the mechanism by which the triple bond is broken is important, because the scission of small molecules containing multiple bonds between the atoms is often the pivotal step in many catalytic processes.

The formation of carbide atoms in clusters often results from the dissociation of CO with subsequent (or concomitant) loss of the oxygen as $CO_2$. There are many reactions known that generate carbide atoms and a few examples involving hexaruthenium clusters will be given here as these have been studied in detail. Using $^{13}C$ labelling experiments, the carbide atom in $Ru_6C(CO)_{17}$ was shown to originate from the disproportionation of a coordinated CO ligand according to the equation shown below:

$$2\ CO \rightarrow\ 'C'\ +\ CO_2$$

The mechanism by which the $C\equiv O$ bond is broken involves an intermediate in which both the carbon and oxygen atoms of the carbonyl group are coordinated to the cluster skeleton. This 'side-on' $\eta^2$-coordination considerably weakens the $C\equiv O$ bond so that it is prone to complete scission.

Heating $[Ru_6(CO)_{18}]^{2-}$ in refluxing diglyme for one hour affords the carbide cluster $[Ru_6C(CO)_{16}]^{2-}$ in high yield. The same cluster $[Ru_6(CO)_{18}]^{2-}$ also undergoes conversion to the neutral carbide cluster $Ru_6C(CO)_{17}$ on reaction with trifluoroethanoic anhydride (see Scheme 7.16).

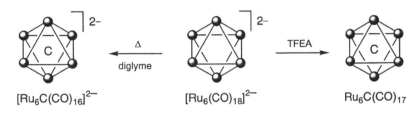

$[Ru_6C(CO)_{16}]^{2-}$        $[Ru_6(CO)_{18}]^{2-}$        $Ru_6C(CO)_{17}$

SCHEME 7.16  CARBIDE FORMATION USING THERMAL AND CHEMICAL METHODS.

The above reactions do not provide any mechanistic information about the $C\equiv O$ bond cleavage. However, two reactions that do provide some mechanistic insights are described below. The thermolysis of $Ru_3(CO)_{12}$ in heptane containing trimethylbenzene affords several products including $Ru_6(\eta^2-\mu_4-CO)_2(CO)_{13}(\eta-C_6H_3Me_3-1,3,5)$. This cluster has a bi-edged bridged tetrahedral metal core and is shown in Scheme 7.17. The two edge bridges form *pseudo*-butterfly cavities with the tetrahedron and within each cavity there is a $\eta^2$-CO ligand. Heating $Ru_6(\eta^2-\mu_4-CO)_2(CO)_{13}(\eta-C_6H_3Me_3-1,3,5)$ brings about the conversion to the carbide containing cluster $Ru_6C(CO)_{14}(\eta-C_6H_3Me_3-1,3,5)$ together with $HRu_6(\eta^2-\mu_4-CO)(CO)_{13}(\mu_2-C_6H_3Me_2CH_2-1,3,5)$ (in equal amounts), and carbon dioxide is also evolved.

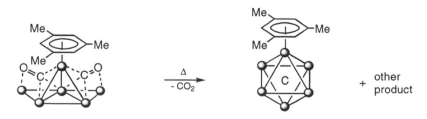

$Ru_6(\eta-\mu_4-CO)_2(CO)_{13}(\eta-C_6H_3Me_3-1,3,5)$      $Ru_6C(CO)_{14}(\eta-C_6H_3Me_3-1,3,5)$

SCHEME 7.17  CARBIDE FORMATION VIA A $\eta^2$-CO LIGAND.

From these observations it is believed that the formation of the carbide atom involves the cleavage of a $\eta^2$-CO ligand via nucleophilic attack of the oxygen on a terminal CO carbon of a second cluster molecule. Elimination of $CO_2$ from this intermediate would generate the carbide atom. Subsequent rearrangement of the metal polyhedron would lead to the formation of $Ru_6C(CO)_{14}(\eta\text{-}C_6H_3Me_3\text{-}1,3,5)$.

An intriguing cluster that reacts to give a carbide atom located inside an octahedral ruthenium cavity is $Ru_6(\mu_5\text{-}C)(CO)_{15}(\mu_3\text{-}C_{16}H_{16}\text{-}\mu\text{-}O)$ (see Scheme 7.18). The metal polyhedron in $Ru_6(\mu_5\text{-}C)(CO)_{15}(\mu_3\text{-}C_{16}H_{16}\text{-}\mu\text{-}O)$ is relatively open with only nine Ru-Ru bonds in comparison to the twelve metal-metal bonds in an octahedron. A carbide atom occupies the central cavity and interacts with five of the six ruthenium atoms. An oxo atom spans two ruthenium atoms and also bonds to the organic ring. It is thought that the carbide and oxo atoms are both derived from the same CO ligand, and represents an intermediate between a coordinated $\eta^2$-CO ligand and a $\mu_6$-C atom. The pyrolysis of $Ru_6C(CO)_{15}(\mu_3\text{-}C_{16}H_{16}\text{-}\mu\text{-}O)$ affords the octahedral carbide containing cluster $Ru_6C(CO)_{14}(\mu_3\text{-}C_{16}H_{16})$ and $CO_2$ in quantitative yield.

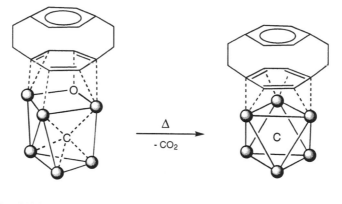

$$Ru_6C(CO)_{15}(\mu_3\text{-}C_{16}H_{16}\text{-}\mu\text{-}O) \qquad\qquad Ru_6C(CO)_{14}(\mu_3\text{-}C_{16}H_{16})$$

SCHEME 7.18  CARBIDE FORMATION BY LOSS OF $CO_2$.

## 7.7 CLUSTER DECOMPOSITION REACTIONS

Most of the reactions described in this chapter do not involve a change in the number of metal atoms present in the cluster. In Chapter 6 the majority of cluster reactions involved increasing the number of metal atoms present. In contrast, cluster decomposition reactions are those which result in the nuclearity of the cluster decreasing. They are sometimes referred to as degradation reactions.

An example of a decomposition reaction is the breakdown of $[Ru_{10}C(CO)_{24}]^{2-}$ to $[Ru_6C(CO)_{16}]^{2-}$ and $Ru_3(CO)_{12}$ under a stream of CO at atmospheric pressure as shown in Scheme 7.19. However, both $[Ru_6C(CO)_{16}]^{2-}$ and $Ru_3(CO)_{12}$ can be made by much more convenient and higher yielding routes. A decomposition reaction that gives a product unavailable by other routes is the preparation of the square pyramidal cluster $Ru_5C(CO)_{15}$ from $Ru_6C(CO)_{17}$ in a degradative carbonylation reaction (see Scheme 7.19). A vertex is removed from the octahedral cluster under a high pressure of carbon monoxide to give $Ru_5C(CO)_{15}$ and $Ru(CO)_5$. This is the only known method for preparing $Ru_5C(CO)_{15}$ and thus illustrates the important synthetic utility of the decomposition reaction.

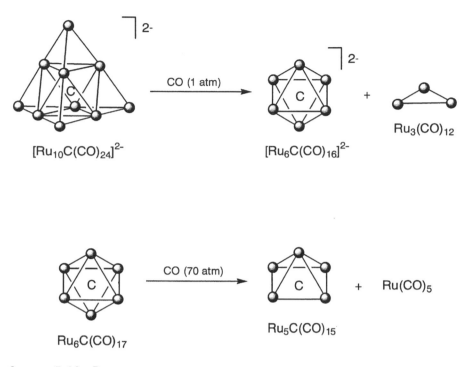

$[Ru_{10}C(CO)_{24}]^{2-}$

CO (1 atm)

$[Ru_6C(CO)_{16}]^{2-}$

$+$

$Ru_3(CO)_{12}$

$Ru_6C(CO)_{17}$

CO (70 atm)

$Ru_5C(CO)_{15}$

$+$

$Ru(CO)_5$

SCHEME 7.19 DEGRADATIVE CARBONYLATION REACTIONS.

CHAPTER 8

# REACTIONS WITH ORGANIC LIGANDS

The reactivity of transition metal carbonyl clusters with unsaturated organic compounds is a particularly important area within the field of cluster chemistry. Many of the reactions that take place are related to organic transformations that are heterogeneously catalysed on metal surfaces. In this chapter the key types of organometallic reactions in cluster chemistry are described. These reactions are generally more complicated than related transformations in mononuclear complexes due to the availability of multicentre bonding sites for the ligands as well as the possibility of changes to the cluster core.

## 8.1 ALKYL AND ALKYLIDYNE CLUSTERS

There are very few examples of transition metal carbonyl clusters that contain alkyl ligands ($CR_3$). However, in the small number of cases that are known, their chemistry hardly differs to those observed in mononuclear complexes with insertion of a carbonyl ligand into the metal-alkyl bond being the main reaction. An example of the synthesis of a cluster-alkyl compound is provided by the reaction of $[Ru_6C(CO)_{16}]^{2-}$ with methyl iodide to give the monoanionic cluster $[Ru_6C(CO)_{16}(Me)]^-$, as shown in Scheme 8.1. Under a high pressure of carbon monoxide $[Ru_6C(CO)_{16}(Me)]^-$ undergoes a carbonyl insertion reaction to afford $[Ru_6C(CO)_{16}(COMe)]^-$.

In contrast to simple alkyl ligands, alkylidyne ligands ($\mu_3$-CR) are much more common in cluster chemistry. The most well documented examples of clusters with

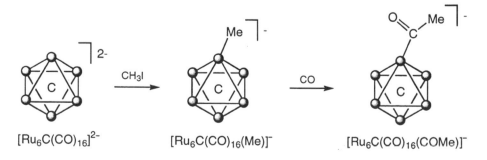

$[Ru_6C(CO)_{16}]^{2-}$       $[Ru_6C(CO)_{16}(Me)]^-$       $[Ru_6C(CO)_{16}(COMe)]^-$

SCHEME 8.1 THE SYNTHESIS OF $[Ru_6C(CO)_{16}(Me)]^-$ AND SUBSEQUENT CARBONYLATION TO $[Ru_6C(CO)_{16}(COMe)]^-$.

alkylidyne ligands are based on tricobalt species with the formula $Co_3(CO)_9(\mu_3\text{-}CR)$ (where R = H, alkyl, halogen, aryl or even another metal group). Their preparation is quite straightforward and involves reacting the dicobalt complex $Co_2(CO)_8$ with $RCX_3$ (X = Cl and Br) as shown in Scheme 8.2. Reactions involving the R-group in these clusters have been extensively studied. For example, $Co_3(CO)_9(\mu_3\text{-}CBr)$ reacts with ethyne ($HC\equiv CH$) in the presence of $Cu^+$ and $EtNH_2$ to form the metalloalkyne compound $[Co_3(CO)_9(\mu_3\text{-}C_2)]_2$. The role of the copper(I) salt is to abstract $Br^-$ from the cluster while the base deprotonates the ethyne. Compounds like the one shown in Scheme 8.2 in which two clusters are connected by an alkyne are of interest for their non-linear optical properties.

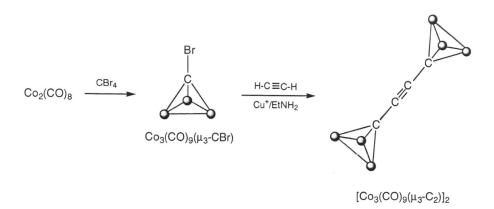

$Co_2(CO)_8$ $\xrightarrow{CBr_4}$ $Co_3(CO)_9(\mu_3\text{-}CBr)$ $\xrightarrow[Cu^+/EtNH_2]{H\text{-}C\equiv C\text{-}H}$ $[Co_3(CO)_9(\mu_3\text{-}C_2)]_2$

SCHEME 8.2 THE PREPARATION OF THE ALKYLIDYNE CLUSTER $Co_3(CO)_9(\mu_3\text{-}CBr)$ AND SUBSEQUENT COUPLING WITH AN ALKYNE TO GIVE $[Co_3(CO)_9(\mu_3\text{-}C_2)]_2$.

Alternatively, the reaction of $Co_3(CO)_9(\mu_3\text{-}CBr)$ with $[Ru(\eta\text{-}C_5H_5)(CO)_2]^-$ and $[HgRu(\eta\text{-}C_5H_5)(CO)_2]^+$ affords the mixed-metal cluster $Co_4Ru_2HgC(CO)_{13}(\eta\text{-}C_5H_5)_2$ (see Scheme 8.3). This reaction is clearly quite complicated and involves, to some extent, fragmentation of the precursor cluster compound, which then reassembles with the other two reagents. In order to prevent the core of the tricobalt cluster from fragmenting during reaction, one carbonyl ligand on each cobalt can be substituted beforehand by a chelating face-capping *tris*-phosphine ligand such as *tris*-(diphenylphosphino)methane (tdpm) as shown in the Scheme. The tricobalt core in the resulting product, $Co_3(CO)_6(tdpm)(\mu_3\text{-}CBr)$, is considerably more stable and in the reaction with the ruthenate species $[Ru(\eta\text{-}C_5H_5)(CO)_2]^-$ remains intact to afford $Co_3(CO)_6(tdpm)\{\mu_3\text{-}C\text{-}Ru(\eta\text{-}C_5H_5)(CO)_2\}$. This cluster represents an example of a tetrahedral $sp^3$ carbon atom surrounded solely by metal atoms.

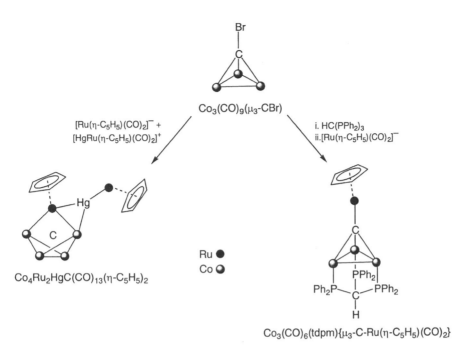

$Co_3(CO)_9(\mu_3\text{-CBr})$

$[Ru(\eta\text{-}C_5H_5)(CO)_2]^-$ +
$[HgRu(\eta\text{-}C_5H_5)(CO)_2]^+$

i. $HC(PPh_2)_3$
ii. $[Ru(\eta\text{-}C_5H_5)(CO)_2]^-$

$Co_4Ru_2HgC(CO)_{13}(\eta\text{-}C_5H_5)_2$

Ru ●
Co ◯

$Co_3(CO)_6(tdpm)\{\mu_3\text{-}C\text{-}Ru(\eta\text{-}C_5H_5)(CO)_2\}$

SCHEME 8.3  REACTIONS OF $Co_3(CO)_9(\mu_3\text{-CBr})$ INVOLVING BOTH FRAGMENTATION AND RETENTION OF THE $Co_3C$ UNIT.

## Syntheses of $Co_4(\mu_4\text{-GeMe})_2(CO)_{11}$

The alkylidyne (CR) ligand prefers to bond to a trimetal face to maintain sp³ hybridisation at the carbon atom. However, the larger germanium (GeR) ligand often coordinates over a square/rectangular face ($\mu_4$-coordination). This is illustrated by the ease with which the cluster $Co_4(\mu_4\text{-GeMe})_2(CO)_{11}$ can be isolated from a number of different reactions, summarised in the Scheme. On the whole, the yields

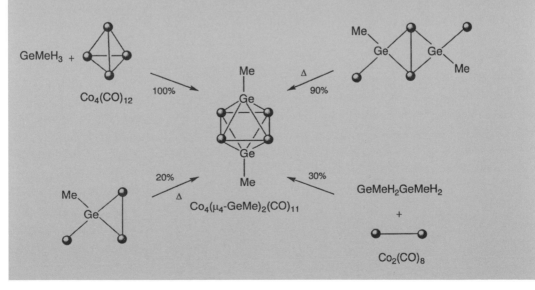

$GeMeH_3$ +

$Co_4(CO)_{12}$  100%

90%  $\Delta$

20%  $\Delta$  $Co_4(\mu_4\text{-GeMe})_2(CO)_{11}$  30%  $GeMeH_2GeMeH_2$

+

$Co_2(CO)_8$

for these reactions are good, and the simplest method involves the direct reaction between $Co_4(CO)_{12}$ and $GeMeH_3$. In addition, these reactions illustrate the way in which certain clusters seem to be thermodynamically favoured, i.e. simply heating the various cobalt compounds always results in the formation of the *pseudo-octahedral* cluster $Co_4(\mu_4\text{-GeMe})_2(CO)_{11}$. Other clusters have been noted to often crop up during thermal reactions and they are frequently based on octahedral structures. This implies that the octahedron is a particular stable geometry in cluster chemistry.

## 8.2   ALKENES, ALKYNES AND ALLYLS

Simple alkene complexes of clusters are used as labile groups in much the same way as they are used in mononuclear chemistry. Alkenes such as ethene tend to be very labile and may be displaced by a range of other ligands. Alkene derivatives of the tetrairidium cluster $Ir_4(CO)_{12}$ are useful synthetic intermediates, as this cluster is otherwise particularly difficult to activate. The reaction of $[Ir_4(CO)_{11}Br]^-$ with an alkene and $AgBF_4$ affords $Ir_4(CO)_{11}(\eta^2\text{-alkene})$ (alkene = $C_2H_4$, $C_3H_6$, cyclooctadiene or norbornadiene) as shown in Scheme 8.4. The $AgBF_4$ abstracts bromide from the cluster generating a vacant coordination site to which the alkene coordinates. These alkene derivatives are only stable when an excess of the alkene is present.

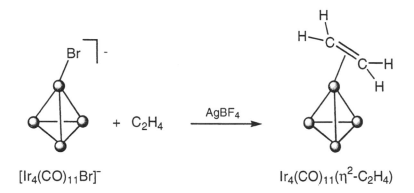

SCHEME 8.4   THE SYNTHESIS OF A CLUSTER WITH A $\eta^2$-ALKENE LIGAND.

Cyclic dienes such as tetraphenylcyclobutadiene are introduced into clusters in a quite unusual ligand transfer reaction. For example, the reaction between the dianionic cluster $[Ru_5C(CO)_{14}]^{2-}$ and the mononuclear dication $[Pd(\eta\text{-}C_4Ph_4)(NCMe)_2]^{2+}$ affords the neutral cluster $Ru_5C(CO)_{13}(\eta\text{-}C_4Ph_4)$. This reaction is illustrated in Scheme 8.5 and shows an intermediate, believed to consist of a heteronuclear cluster in which the palladium caps the $Ru_5C$ skeleton prior to the transfer of the $C_4Ph_4$ ring from the palladium centre to one of the ruthenium atoms.

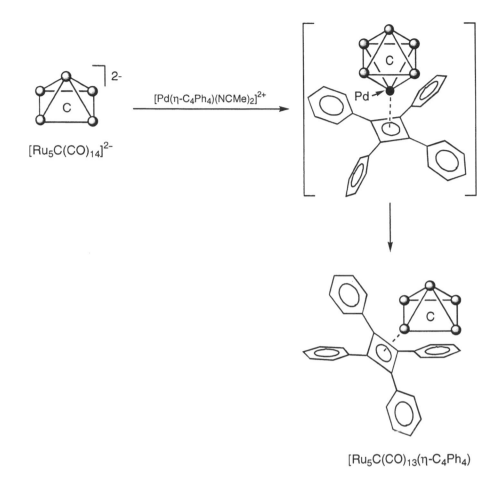

$[Ru_5C(CO)_{14}]^{2-}$

$[Pd(\eta-C_4Ph_4)(NCMe)_2]^{2+}$

$[Ru_5C(CO)_{13}(\eta-C_4Ph_4)]$

SCHEME 8.5 A REDOX MEDIATED LIGAND TRANSFER REACTION THAT PROCEEDS VIA A HETERONUCLEAR INTERMEDIATE.

The synthesis and reactivity of cluster-alkyne compounds has been extensively studied. Consider the reaction between the activated triosmium cluster $Os_3(CO)_{10}(NCMe)_2$ and a terminal alkyne $HC{\equiv}CR$. The initial product from this reaction is $Os_3(CO)_{10}(\mu_3\text{-HCCR})$, in which the alkyne bonds over the $Os_3$ face providing four electrons towards cluster bonding (see Scheme 8.6). On further heating a CO ligand is expelled, which induces the terminal hydrogen atom to transfer from the alkyne to the cluster. In addition, the carbon to which the hydrogen atom was attached forms a $\sigma$-bond with the cluster to yield $HOs_3(CO)_9(\mu_3\text{-CCR})$, and the ligand formally donates five electrons to the cluster. Depending upon the nature of the R-group and the actual reagent added, several different reactions of $HOs_3(CO)_9(\mu_3\text{-CCR})$ can take place. For example, treatment of $HOs_3(CO)_9(\mu_3\text{-CCR})$ with LiBHEt$_3$ affords two products. First, $H_2Os_3(CO)_9(\mu_3\text{-HCCR})$ in which the alkyne has been regenerated, and second, $HOs_3(CO)_9(\mu_3\text{-C}_3H_2R)$ which contains a face-capping allyl derivative. The face-capping bonding mode displayed by

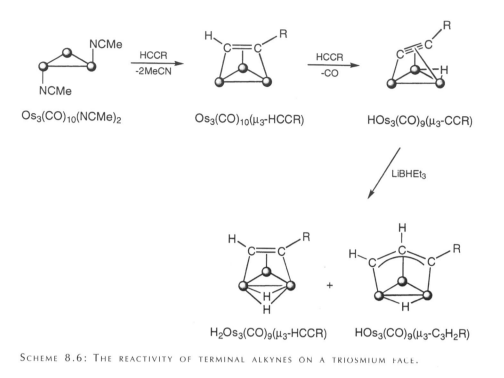

SCHEME 8.6: THE REACTIVITY OF TERMINAL ALKYNES ON A TRIOSMIUM FACE.

the allyl ligand in HOs$_3$(CO)$_9$($\mu_3$-C$_3$H$_2$R) is typical of this type of ligand in clusters, prefered to simple $\eta^3$ coordination observed at single metal centres.

Clusters can induce alkynes to undergo transformations to ligands other than allyls. A relatively common reaction that is observed in both mononuclear organo-metallic chemistry and cluster chemistry is the cyclisation of alkynes to form coordinated rings, such as C$_4$R$_4$ and C$_6$R$_6$. With clusters, an even more remarkable reaction can take place, the cyclisation of alkynes to form five-membered rings. This is a much more complicated process and not only involves the coupling of the alkynes, but also breaking of a C≡C triple bond (see Scheme 8.7).

The reaction of Ru$_6$C(CO)$_{17}$ with phenylacetylene (HC≡CPh) and two equivalents of trimethylamine *N*-oxide (Me$_3$NO, used to activate the cluster by removing coordinated CO as CO$_2$ and thereby creating, in this case, two vacant coordination sites on the surface of the cluster, see Chapter 6) gives the expected product Ru$_6$C(CO)$_{15}$($\mu_3$-HCCPh). Reaction of this cluster with a further equivalent of HC≡CPh and Me$_3$NO affords Ru$_6$C(CO)$_{14}${$\mu_3$-C(Ph)CHC(Ph)CH}, which contains a 1,3-diene ligand. The 1,3-diene ligand is produced from the insertion of the incoming alkyne into the Ru-C bond of the alkyne already coordinated. This cluster subsequently undergoes conversion to Ru$_6$C(CO)$_{13}$($\eta$-C$_5$H$_3$Ph$_2$-1,3)($\mu_3$-CPh) when reacted with further HC≡CPh, and Me$_3$NO. In this final step, the scission of a C≡C bond takes place forming a 1,3-diphenylcyclopentadienyl ring. The other fragment of the cleaved alkyne gives rise to an alkylidyne ligand that is bonded over one of the faces in the same cluster.

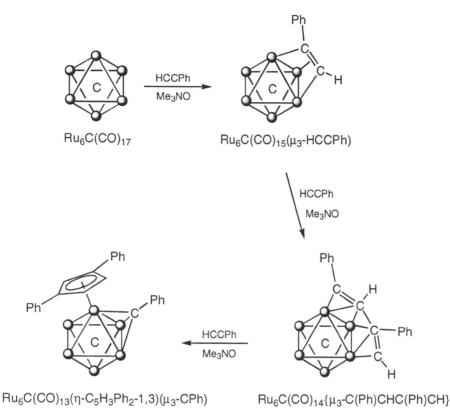

$Ru_6C(CO)_{17}$        $Ru_6C(CO)_{15}(\mu_3\text{-HCCPh})$

$Ru_6C(CO)_{13}(\eta\text{-}C_5H_3Ph_2\text{-}1,3)(\mu_3\text{-CPh})$      $Ru_6C(CO)_{14}\{\mu_3\text{-C(Ph)CHC(Ph)CH}\}$

SCHEME 8.7 THE REACTIONS OF $Ru_6C(CO)_{17}$ WITH PHENYLACETYLENE LEADING TO THE FORMATION OF A $C_5$-RING.

## 8.3 C₅-RINGS

Cyclopentadienyl complexes are amongst the most important in mononuclear organometallic chemistry. Cyclopentadienyl ligands have also been studied in some detail in cluster chemistry, but in general, they do not display the same diversity as that observed in their mononuclear counterparts. In the previous section the formation of a 1,3-substituted cyclopentadienyl ring, from alkyne ligands on the face of a cluster, was described. However, this type of reaction is extremely rare and more general and direct routes can be used to introduce cyclopentadienyl ligands into clusters. One strategy involves using ionic coupling reactions between fragments such as $[Rh(\eta\text{-}C_5H_5)(NCMe)_3]^{2+}$ and $[Ru(\eta\text{-}C_5H_5)(NCMe)_3]^+$ with appropriate cluster anions. Some examples of these reactions are illustrated in Scheme 8.8. The reactions shown in the Scheme make use of the capping fragment $[Ru(\eta\text{-}C_5H_5)(NCMe)_3]^+$, which is a monocation and therefore reacts with monoanions to form mono-capped clusters and with dianions to form bi-capped clusters. The three acetonitrile ligands (MeCN) are weakly coordinated to the ruthenium fragment and are readily lost in the capping reaction in preference to the chelating cyclopentadienyl group.

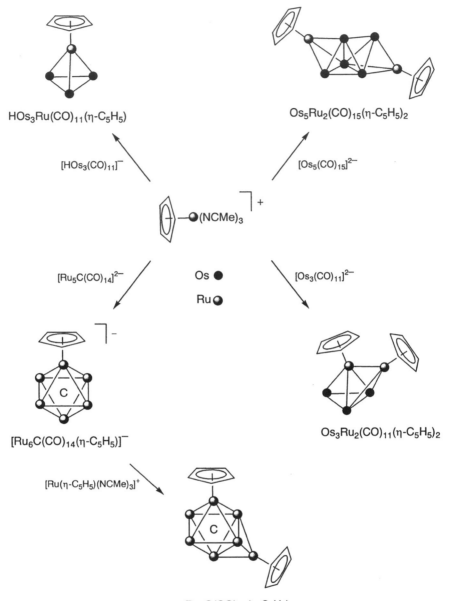

$HOs_3Ru(CO)_{11}(\eta\text{-}C_5H_5)$

$Os_5Ru_2(CO)_{15}(\eta\text{-}C_5H_5)_2$

$[HOs_3(CO)_{11}]^-$

$[Os_5(CO)_{15}]^{2-}$

$[Ru_5C(CO)_{14}]^{2-}$

$[Os_3(CO)_{11}]^{2-}$

Os ●

Ru ◐

$[Ru_6C(CO)_{14}(\eta\text{-}C_5H_5)]^-$

$Os_3Ru_2(CO)_{11}(\eta\text{-}C_5H_5)_2$

$[Ru(\eta\text{-}C_5H_5)(NCMe)_3]^+$

$Ru_7C(CO)_{14}(\eta\text{-}C_5H_5)_2$

SCHEME 8.8 SOME REDOX CONDENSATION REACTIONS USING THE FRAGMENT $[Ru(\eta\text{-}C_5H_5)(NCMe)_3]^+$.

The trinuclear cluster $[HOs_3(CO)_{11}]^-$ reacts with $[Ru(\eta\text{-}C_5H_5)(NCMe)_3]^+$ to afford the neutral, tetrahedral mixed-metal cluster $HOs_3Ru(CO)_{11}(\eta\text{-}C_5H_5)$. The osmium dianions $[Os_3(CO)_{11}]^{2-}$ and $[Os_5(CO)_{15}]^{2-}$ react with two equivalents of $[Ru(\eta\text{-}C_5H_5)(NCMe)_3]^+$ to afford $Os_3Ru_2(CO)_{11}(\eta\text{-}C_5H_5)_2$ and $Os_5Ru_2(CO)_{15}$ $(\eta\text{-}C_5H_5)_2$, respectively. The cluster core in $Os_3Ru_2(CO)_{11}(\eta\text{-}C_5H_5)_2$ is a trigonal bipyramid, but not with the three osmium atoms forming the central equatorial triangle. Instead, one ruthenium fragment caps a face and the other bridges an edge and they also bond to each other. In $Os_5Ru_2(CO)_{15}(\eta-C_5H_5)_2$ the trigonal bipyramidal osmium core is capped on two opposite faces to give a cluster that can be viewed as having a polytetrahedral (i.e. a series of four fused tetrahedra in a row) mixed-metal core. The square face of the square pyramidal cluster $[Ru_5C(CO)_{14}]^{2-}$ is capped in preference to one of the triangular faces when it is reacted with the $[Ru(\eta-C_5H_5)(NCMe)_3]^+$ fragment. The resulting monoanionic cluster has an octa-hedral metal framework. Further reaction of this monoanion results in a monocapped octahedron carrying two cyclopentadienyl ligands. Other methodologies that result in heteronuclear clusters are found in Chapter 9.

Cyclopentadienyl ligands may also be introduced into clusters using a variety of other methods. For example, $Na(C_5H_5)$ and cyclopentadiene ($C_5H_6$) are both useful reagents and illustrations of the use of $C_5H_6$ is given in Scheme 8.9. The reaction of the activated cluster $Os_3(CO)_{10}(NCMe)_2$ with cyclopentadiene affords $Os_3(CO)_{10}(\eta^4-C_5H_6)$. In this reaction the two weakly coordinated acetonitrile ligands are replaced by the cyclopentadiene ligand. Conversion of the $C_5H_6$ ligand into a cyclopentadienyl ($C_5H_5$) moiety is achieved by treatment of $Os_3(CO)_{10}(\eta^4-C_5H_6)$ with $[Ph_3C][BF_4]$. This reagent is used to abstract hydrides (i.e. removal of $H^-$ to form $Ph_3CH$). In this case it removes $H^-$ from $C_5H_6$ to produce the cationic cluster $[Os_3(CO)_{10}(\eta-C_5H_5)][BF_4]$. Subsequent reaction of $[Os_3(CO)_{10}(\eta-C_5H_5)]^+$ with a hydride source affords the neutral linear cluster $HOs_3(CO)_{10}(\eta-C_5H_5)$ as shown in the Scheme.

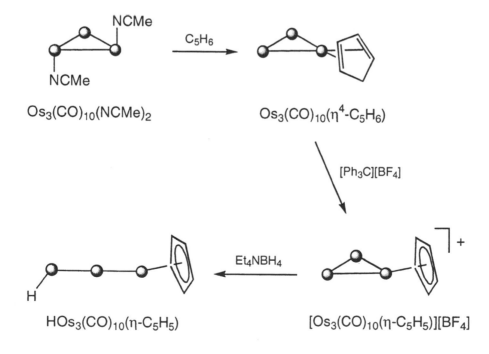

**SCHEME 8.9** THE SYNTHESIS OF CLUSTERS WITH CYCLOPENTADIENYL LIGANDS COMMENCING WITH CYCLOPENTADIENE.

## 8.4   C₆-RINGS

Benzene and other arene ligands can be introduced into clusters in a number of ways including:

i.   Direct thermal reaction between a cluster and an arene.
ii.  Using suitable arene-containing fragments for redox condensation.
iii. Generating the arene from non-aromatic precursors such as cyclohexadienes and alkynes.

Examples of these reactions will be described here.

The electronically unsaturated cluster $H_2Os_3(CO)_{10}$ undergoes reaction with cyclohexadiene ($C_6H_8$) in refluxing octane to afford $HOs_3(CO)_9(\mu_3\text{-}C_6H_7)$ that has a face-capping cyclohexadienyl ligand (see Scheme 8.10). Treatment of $HOs_3(CO)_9(\mu_3\text{-}C_6H_7)$ with the hydride abstractor $[Ph_3C][BF_4]$ affords the cationic cluster $[HOs_3(CO)_9(\mu_3\text{-}C_6H_6)]^+$ in which the hydride has been removed from the organic ring to give a face-capping $C_6H_6$ ligand. The cationic cluster reacts with the non-coordinating base 1,8-diazabicyclo[5.4.0]undeca-7-ene (DBU), which removes the bridging hydride from the cluster as H⁺ to yield the neutral species $Os_3(CO)_9(\mu_3\text{-}C_6H_6)$. Finally, irradiation of $Os_3(CO)_9(\mu_3\text{-}C_6H_6)$ with UV light activates the transfer of two hydrogen atoms from the benzene ring to the cluster to afford $H_2Os_3(CO)_9(\mu_3\text{-}C_6H_4)$, which has a face-capping benzyne ligand. This reaction may be described as a photoinduced isomerisation. The resulting benzyne ligand bonds to the triosmium core in a manner very similar to that of many alkyne ligands, and provides four electrons towards bonding.

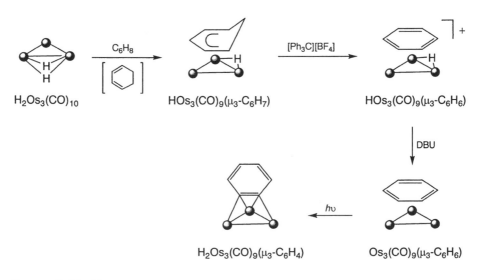

SCHEME 8.10   THE CONVERSION OF CYCLOHEXADIENE TO BENZENE AND THEN BENZYNE ON A TRIOSMIUM CLUSTER FACE.

A slightly different approach used to prepare cluster-benzene compounds, also commencing with cyclohexadiene, involves reaction in the presence of $Me_3NO$, which activates the cluster towards reaction by oxidative removal of CO. An example of this reaction involves treatment of $Ru_5C(CO)_{15}$ with two equivalents of $Me_3NO$ in the presence of excess cyclohexadiene to yield $Ru_5C(CO)_{13}(\mu_2\text{-}C_6H_8)$ (see Scheme 8.11). The cyclohexadiene ligand coordinates across a Ru-Ru edge along the square base of the cluster. Subsequent reaction of $Ru_5C(CO)_{13}(\mu_2\text{-}C_6H_8)$ with another aliquot of $Me_3NO$ brings about the expulsion of one more CO ligand and creation of a vacant coordination site. In order for the cluster to regain coordinative saturation, the cyclohexadiene (a four electron donor) undergoes dehydrogenation to benzene (a six electron donor), which coordinates over a triangular face of the cluster affording $Ru_5C(CO)_{12}(\mu_3\text{-}C_6H_6)$. Heating $Ru_5C(CO)_{12}(\mu_3\text{-}C_6H_6)$ in hexane for a few hours causes the face-capping benzene ligand to migrate to a basal ruthenium atom affording $Ru_5C(CO)_{12}(\eta\text{-}C_6H_6)$ in quantitative yield.

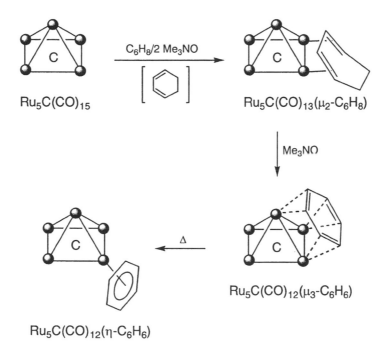

SCHEME 8.11 THE REACTION OF $Ru_5C(CO)_{15}$ WITH CYCLOHEXADIENE.

Buckminsterfullerene ($C_{60}$) coordinates to clusters via a $C_6$-ring in the face-capping bonding mode analogous to the bonding observed for benzene. Preparation of these clusters simply involves heating the appropriate cluster directly with the $C_{60}$ ligand. The structures of two compounds, $[Ru_3(CO)_9]_n(C_{60})$ (where n = 1 and 2), are shown in Figure 8.1.

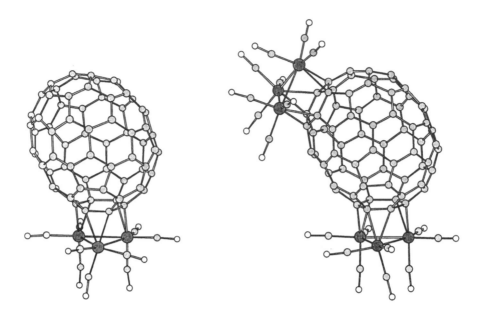

FIGURE 8.1   THE STRUCTURES OF $[Ru_3(CO)_9]_n(C_{60})$ (WHERE $n = 1$ AND $2$).

## 8.5   $C_7$ AND $C_8$-RINGS

The thermal reaction of $Ru_3(CO)_{12}$ with cycloheptatriene ($C_7H_8$) affords several products including $Ru_4(CO)_7(\mu\text{-}C_7H_7)_2$ in low yield. The structure of $Ru_4(CO)_7(\mu\text{-}C_7H_7)_2$ is quite remarkable and is shown in Figure 8.2. It may be described as a type of sandwich compound, in which a $Ru_4$ tetrahedron is sandwiched between two edge-bridging $C_7H_7$ cycloheptatrienyl rings.

It is also possible to prepare cluster-cycloheptatrienyl compounds from $[C_7H_7]^+$ (tropylium). In an unusual redox reaction, $[Ru_6C(CO)_{16}]^{2-}$ reacts with two equiva-

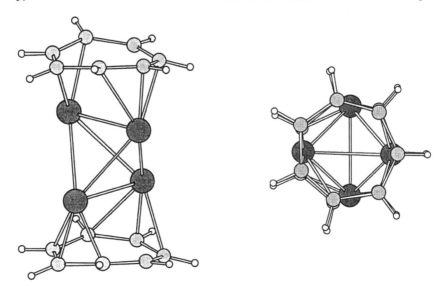

FIGURE 8.2   TWO VIEWS OF THE STRUCTURE OF $Ru_4(CO)_7(\mu\text{-}C_7H_7)_2$.

lents of tropylium bromide, $[C_7H_7]Br$, to yield the neutral cluster with a bitropyl ligand, $Ru_6C(CO)_{14}(\mu_3\text{-}C_7H_6C_7H_6)$ as shown in Scheme 8.12. The two $C_7$-rings have fused with the formation of a C–C single bond and the coordinated cycloheptatrienyl ring bonds over a triruthenium face.

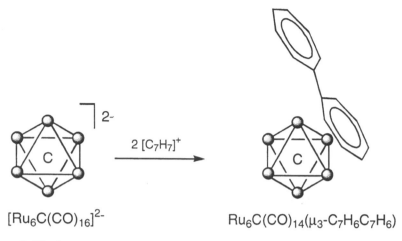

$[Ru_6C(CO)_{16}]^{2-}$                    $Ru_6C(CO)_{14}(\mu_3\text{-}C_7H_6C_7H_6)$

SCHEME 8.12  IONIC COUPLING BETWEEN A CLUSTER ANION AND ORGANIC CATION.

The reactions of clusters with $C_8$-rings are quite diverse, partly due to the range of $C_8$-rings available, e.g. cyclooctene, cycloocta-1,3-diene, cycloocta-1,5-diene, cyclooctatetraene. Several products of varying nuclearity are often isolated from the same reaction. One of the most interesting reactions involving a $C_8$-ring is the reaction between cycloocta-1,3-diene and $Ru_3(CO)_{12}$. Several products containing $C_8$-rings coordinated in various ways are isolated. One of the products is the hexaruthenium cluster $HRu_6(\mu_4\text{-}\eta^2\text{-}CO)_2(CO)_{13}(\eta^5\text{-}C_5H_3C_3H_6)$, which is shown in Figure 8.3.

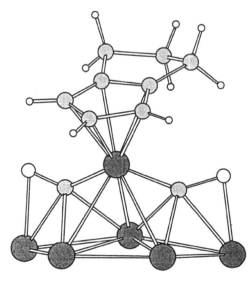

FIGURE 8.3  THE STRUCTURE OF $HRu_6(\mu_4\text{-}\eta^2\text{-}CO)_2(CO)_{13}(\eta^5\text{-}C_5H_3C_3H_6)$.

The cycloocta-1,3-diene has undergone contraction to form two fused $C_5$-rings, one saturated and one unsaturated. This type of ligand is a trihydropentalene and results from the formation of a C-C bond across the $C_8$-ring (as well as hydrogen migration about the ring). Ring contraction is a process that is often catalysed by heterogeneous catalysts. In molecular chemistry, however, it is very rare, although several examples involving clusters are known.

A more selective route used to prepare clusters with coordinated $C_8$-rings involves using preformed metal-ligand fragments, such as the platinum-cycloocta-1,5-diene fragment, $Pt(\eta^4\text{-}C_8H_{12})_2$, which condenses with a number of clusters to form heteronuclear derivatives. Cycloocta-1,5-diene is a widely used ligand in organometallic chemistry and tends to be quite labile. As such, one of the ligands is readily lost from $Pt(\eta^4\text{-}C_8H_{12})_2$ in condensation reactions. For example, the reaction of $Pt(\eta^4\text{-}C_8H_{12})_2$ with the dirhodium complex $Rh_2(CO)_2(\eta\text{-}C_5Me_5)$, which contains a Rh=Rh double bond, affords the trinuclear species, $Rh_2Pt(CO)_2(\eta\text{-}C_5Me_5)_2(\eta^4\text{-}C_8H_{12})$, (see Scheme 8.13).

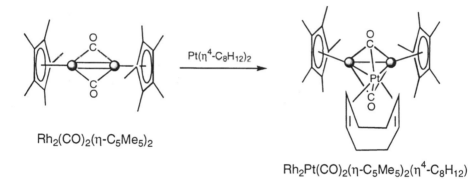

$Rh_2(CO)_2(\eta\text{-}C_5Me_5)_2$

$Rh_2Pt(CO)_2(\eta\text{-}C_5Me_5)_2(\eta^4\text{-}C_8H_{12})$

SCHEME 8.13 A CONDENSATION REACTION INVOLVING AN UNSATURATED Rh=Rh DIMER WITH A PLATINUM FRAGMENT.

## 8.6 HETEROCYCLES

The reactions of clusters with heterocycles have been widely studied, as coordination of the heteroatom at one metal centre and C-H activation at an adjacent metal centre provides novel pathways with applications in organic synthesis. Examples of selective C-H bond activation in heterocycles are described in Chapter 10. For now, a few reactions that illustrate the way in which heterocycles can coordinate to clusters will be given.

In a relatively straightforward example, the activated cluster $H_2Os_3(CO)_{10}$ reacts with pyridine to form $HOs_3(CO)_{10}(\mu\text{-}C_5H_4N)$ (see Scheme 8.14). Loss of hydrogen takes place, so that the pyridine ring bonds to the cluster through both the lone pair on the N-atom and at an *ortho*-carbon atom. In this bonding mode the ligand donates three electrons towards cluster bonding.

The direct reaction between $Ru_3(CO)_{12}$ and 2-anilinopyridine (PhNHPy) affords $HRu_3(CO)_9(\mu_3\text{-}\eta^2\text{-}NPhPy)$ in high yield as shown in Scheme 8.15. The nitrogen

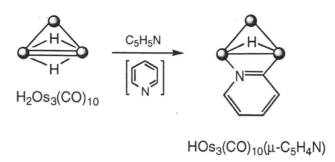

SCHEME 8.14 THE REACTION OF THE UNSATURATED CLUSTER $H_2Os_3(CO)_{10}$ WITH PYRIDINE.

atom of the pyridine ring coordinates to one ruthenium atom while the aniline based nitrogen atom has lost a proton (transferred to the cluster) and bonds across two ruthenium centres. The edge spanned by this nitrogen atom is also bridged by the hydride ligand. The related compound $[Ru_3(CO)_{10}(\mu-\eta^2-NPhPy)]^-$ rapidly forms in high yield from the reaction of $Ru_3(CO)_{12}$ with K[PhNHPy]. In this compound both the nitrogen atoms only bond to two of the three metals.

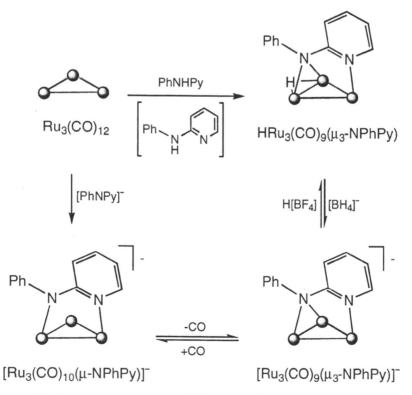

SCHEME 8.15 THE REACTIONS OF $Ru_3(CO)_{12}$ WITH 2-ANILINOPYRIDINE.

Thermal elimination of one CO ligand from $[Ru_3(CO)_{10}(\mu\text{-}\eta^2\text{-NPhPy})]^-$ results in a reorganisation of the 2-anilinopyridine-derived ligand to form the face-capping derivative $[Ru_3(CO)_9(\mu_3\text{-}\eta^2\text{-NPhPy})]^-$ in which the bonding of the ligand resembles that observed in the neutral cluster $HRu_3(CO)_9(\mu_3\text{-}\eta^2\text{-NPhPy})$. This reaction is reversible, and carbonylation results in the regeneration of $[Ru_3(CO)_{10}(\mu\text{-}\eta^2\text{-NPhPy})]^-$. The anionic cluster $[Ru_3(CO)_9(\mu_3\text{-}\eta^2\text{-NPhPy})]^-$ can also be made from the neutral cluster $HRu_3(CO)_9(\mu_3\text{-}\eta^2\text{-NPhPy})$ by deprotonation of the of the metal core using $[BH_4]^-$. Addition of an acid such as $H[BF_4]$ regenerates the neutral cluster $HRu_3(CO)_9(\mu_3\text{-}\eta^2\text{-NPhPy})$ by protonating the metal core.

## Macrocyclic ligands

Macrocyclic ligands can form many interesting multicentred coordination modes with clusters. An example is 1,3,5-trithiane, which adopts a face-capping bonding mode in $Ru_3(CO)_9(\mu_3\text{-}S_3C_3H_6)$ (see Figure). The cluster is isolated in about 50% yield from the reaction between $Ru_3(CO)_{12}$ and 1,3,5-trithiane in refluxing hexane for 2 hours. The 1,3,5-trithiane acts as a 6-electron donor with each sulfur atom eclipsing a ruthenium atom. The macrocycle lies almost parallel to the $Ru_3$ triangle. A symmetric $\mu$-CO bridges each Ru-Ru edge in the metal plane and the remaining CO ligands are all terminal, one axial and one equatorial on each ruthenium atom.

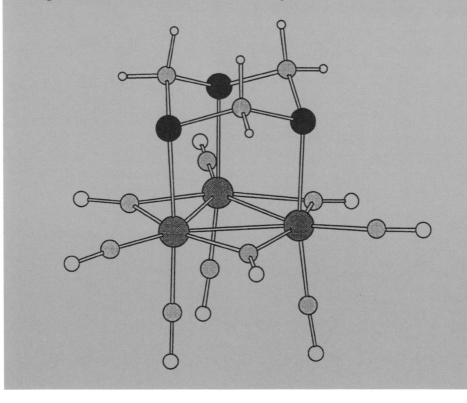

# CHAPTER 9

# HETERONUCLEAR CARBONYL CLUSTERS

Clusters containing two or more different types of metals are termed heteronuclear and have been presented so far in this book without comment. However, they are interesting in their own right not only for their unique properties by virtue of containing polar metal–metal bonds, but also because some of the most elegant examples of designed synthesis in organometallic chemistry have been developed for mixed-metal clusters. This chapter examines both the preparation of heteronuclear clusters and the way in which their reactivity is determined by the mixed-metal bonds.

## 9.1 SYNTHESIS

### 9.1.1 Condensation reactions

Heating small clusters to form larger clusters (pyrolysis or thermolysis) is a widely used method for preparing high nuclearity clusters (see Chapter 6). The driving force behind this process is ligand loss. The thermal elimination of CO or some other two electron ligand induces metal–metal bond formation. While the method is not commonly used for the synthesis of heteronuclear clusters, due to the very complex mixture of products often formed, there are some interesting examples.

An example of cluster condensation is the reaction of $H_4Ru_4(CO)_{12}$ with $Pt_2Ru_4(CO)_{18}$ to form the decanuclear cluster $H_2Pt_2Ru_8(CO)_{23}$ (see Scheme 9.1). The structure of $H_2Pt_2Ru_8(CO)_{23}$ consists of two condensed octahedra that share a Pt–Pt edge. The apices of each octahedron are also connected by metal–metal bonds.

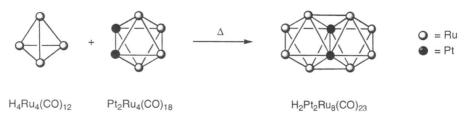

| $H_4Ru_4(CO)_{12}$ | $Pt_2Ru_4(CO)_{18}$ | | $H_2Pt_2Ru_8(CO)_{23}$ | $\bigcirc$ = Ru $\quad$ $\bullet$ = Pt |

SCHEME 9.1 AN EXAMPLE OF A CONDENSATION REACTION.

Reactions like this which involve the addition of the reagents to form a product with the sum of the metal compositions of the precursors, i.e. $Ru_4 + Pt_2Ru_4 = Pt_2Ru_8$, are rare. A more effective route is to condense two compounds that have weakly coordinated ligands that are readily lost during reaction allowing the formation of mixed-metal bonds. For example, the platinum complex $Pt(\eta^4\text{-}C_8H_{12}\text{-}1,5)_2$ reacts with a range of activated osmium complexes to form heteronuclear osmium-platinum clusters, as illustrated in Scheme 9.2. The mononuclear species $Pt(\eta^4\text{-}C_8H_{12}\text{-}1,5)_2$ and $Os(CO)_5$ react to form a planar polytriangular (four triangles fused together) structure as shown in the Scheme. The two triangular clusters $Os_3(CO)_{10}(NCMe)_2$ and $H_2Os_3(CO)_{10}$ react with two equivalents of $Pt(\eta^4\text{-}C_8H_{12}\text{-}1,5)_2$ to afford $Os_3Pt_2(CO)_{10}(\eta^4\text{-}C_8H_{12}\text{-}1,5)_2$ and $H_2Os_3Pt_2(CO)_9(\eta^4\text{-}C_8H_{12}\text{-}1,5)_2$, respectively. In $Os_3Pt_2(CO)_{10}(\eta^4\text{-}C_8H_{12}\text{-}1,5)_2$ the two platinum fragments bond across the same Os–Os edge as well as bonding to each other. In contrast, one of the platinum fragments in $H_2Os_3Pt_2(CO)_9(\eta^4\text{-}C_8H_{12}\text{-}1,5)_2$ bonds over a triosmium face whereas the other bonds across an edge. The structure of $Os_6(CO)_{16}(NCMe)_2$ undergoes a more complicated transformation on reaction with two equivalents of $Pt(\eta^4\text{-}C_8H_{12}\text{-}1,5)_2$. Apart from the addition of two face-capping platinum fragments one of the Os–Os bonds in the original $Os_6$ polyhedron has broken.

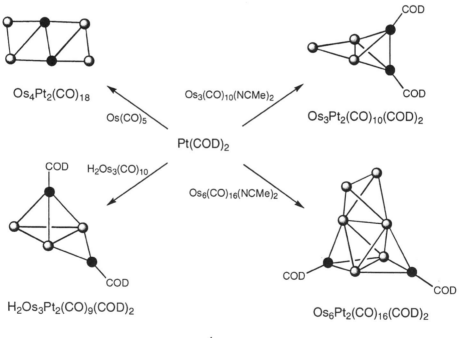

Os$_4$Pt$_2$(CO)$_{18}$

Os(CO)$_5$

Os$_3$(CO)$_{10}$(NCMe)$_2$

COD

COD

Os$_3$Pt$_2$(CO)$_{10}$(COD)$_2$

Pt(COD)$_2$

COD     H$_2$Os$_3$(CO)$_{10}$

Os$_6$(CO)$_{16}$(NCMe)$_2$

COD

H$_2$Os$_3$Pt$_2$(CO)$_9$(COD)$_2$

COD

COD

Os$_6$Pt$_2$(CO)$_{16}$(COD)$_2$

COD = $\eta^4\text{-}C_8H_{12}\text{-}1,5$

SCHEME 9.2   SOME CONDENSATION REACTIONS BETWEEN $Pt(\eta^4\text{-}C_8H_{12}\text{-}1,5)_2$ AND OSMIUM SPECIES UTILISING LABILE GROUPS.

Some of the resulting mixed-metal clusters themselves have weakly coordinated cyclooctadiene ligands attached, which can be replaced by other ligands such as CO or heated to induce further condensation with the formation of even larger clusters.

### 9.1.2 Ligand displacement

Metal complexes that contain a group that is readily lost will react with metal-containing nucleophiles to form new metal–metal bonds. Halide displacement reactions are the most common and have been used to prepare mixed-metal clusters ranging in size from simple dimers to high nuclearity clusters. Ligand displacement reactions can lead to complicated mixtures of products. A particularly notable example is the reaction of $[Ni_6(CO)_{12}]^{2-}$ with $[PtCl_4]^{2-}$ in acetonitrile, which gives many products that are separated by fractional crystallisation. The clusters isolated from this reaction include $[Ni_9(CO)_{18}]^{2-}$, $[Ni_9Pt_3(CO)_{21}]^{4-}$, $[HNi_9Pt_3(CO)_{21}]^{3-}$ $[Ni_{36}Pt_4(CO)_{45}]^{6-}$, $[Ni_{37}Pt_4(CO)_{46}]^{6-}$, $[Ni_{38}Pt_6(CO)_{48}]^{5-}$ and $[HNi_{38}Pt_6(CO)_{48}]^{4-}$. The structures of the metal cores of two of these clusters are shown in Figure 9.1.

The clusters shown in Figure 9.1 are described as having a 'cherry' structure. In $[HNi_{38}Pt_6(CO)_{48}]^{5-}$, a giant octahedral layer of nickel atoms surrounds the $Pt_6$ octahedral core. Similarly, in $[Ni_{37}Pt_4(CO)_{46}]^{6-}$, a central $Pt_4$ tetrahedral core is encapsulated by an incomplete tetrahedron of nickel atoms.

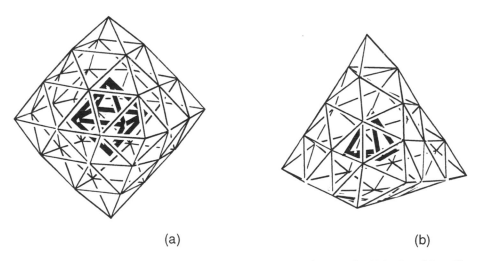

(a)                         (b)

FIGURE 9.1   THE STRUCTURES OF (a) $[HNi_{38}Pt_6(CO)_{48}]^{4-}$ AND (b) $[Ni_{37}Pt_4(CO)_{46}]^{6-}$ SHOWING THE METAL CORES.

Halide displacement also works well for preparing mixed transition-main group metal clusters. For example, the macrocyclic osmium-tin cluster $Os_6Sn_6(CO)_{24}Ph_{12}$ illustrated in Figure 9.2 was prepared by reacting together $Na_2[Os(CO)_4]$ and $Ph_2SnCl_2$. The driving force for the reaction is the formation of NaCl, and is an example of an organometallic metathesis reaction.

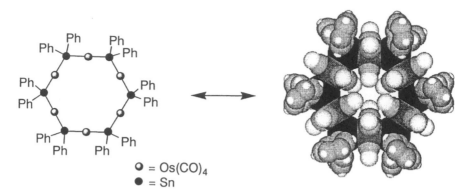

$O$ = Os(CO)$_4$
● = Sn

FIGURE 9.2: TWO REPRESENTATIONS OF THE STRUCTURE OF Os$_6$Sn$_6$(CO)$_{24}$Ph$_{12}$.

## Mercury 'glue'

The addition of mercury salts to anionic metal complexes or clusters often results in the formation of larger clusters. Mercury usually forms the core of the resulting heteronuclear cluster, holding the metallic subunits together like a 'glue'. Some examples are shown below.

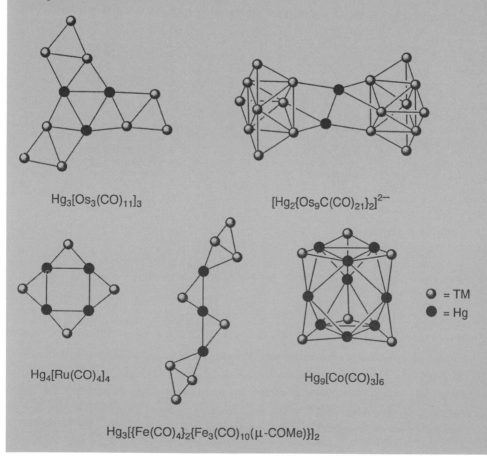

Hg$_3$[Os$_3$(CO)$_{11}$]$_3$

[Hg$_2${Os$_9$C(CO)$_{21}$}$_2$]$^{2-}$

Hg$_4$[Ru(CO)$_4$]$_4$

Hg$_9$[Co(CO)$_3$]$_6$

$O$ = TM
● = Hg

Hg$_3$[{Fe(CO)$_4$}$_2${Fe$_3$(CO)$_{10}$(μ-COMe)}]$_2$

### 9.1.3 Addition reactions

Addition reactions in cluster chemistry may take place when a cluster contains a metal–metal multiple bond. These compounds can accept a metal fragment without the elimination of a ligand, though the metal fragment being added must itself be unsaturated or possess labile ligands. Examples include the reaction of the unsaturated 46-electron cluster $H_2Os_3(CO)_{10}$ with $Co(CO)_2(\eta\text{-}C_5H_5)$ to afford $H_2Os_3Co(CO)_{10}(\eta\text{-}C_5H_5)$ and the reaction of the unsaturated 32-electron dimer $H_2Re_2(CO)_8$ with $Pt(\eta^4\text{-}C_8H_{12}\text{-}1,5)_2$ to form $H_2Re_2Pt(CO)_8(\eta^4\text{-}C_8H_{12}\text{-}1,5)$. Both the unsaturated species contain a metal–metal double bond and the reactions are illustrated in Scheme 9.3.

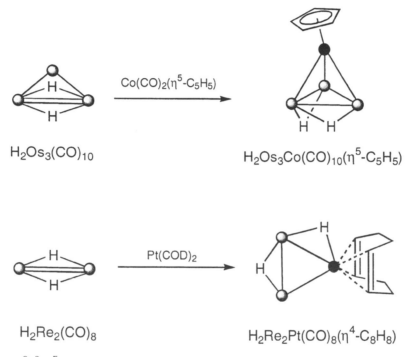

SCHEME 9.3 EXAMPLES OF ADDITION REACTIONS HARNESSING UNSATURATED METAL–METAL BONDS TO AFFORD HETERONUCLEAR CLUSTERS.

Alkylidyne complexes contain a M≡C triple bond that will add metal fragments to generate clusters with triply bridging alkylidyne ($\mu_3$-CR) ligands. Dimers add to alkylidyne complexes to form trinuclear mixed metal clusters. For example, $W(CO)_2(\eta\text{-}C_5H_5)(\equiv CTol)$ reacts with $Co_2(CO)_8$ to give $WCo_2(CO)_8(\eta\text{-}C_5H_5)(\mu_3\text{-}CTol)$ as shown in Scheme 9.4. Mononuclear reagents can be added stepwise, first forming a dinuclear species containing a bridging alkylidyne ligand. In the second addition clusters containing three different metal atoms can be produced. The example shown in Scheme 9.4 involves reacting $W(CO)_2(\eta\text{-}C_5H_5)(\equiv CTol)$ with $Rh(CO)_2(\eta^5\text{-}C_9H_7)$ to produce $WRh(CO)_3(\eta\text{-}C_5H_5)(\mu\text{-}CTol)(\eta^5\text{-}C_9H_7)$, which can then be reacted with $Ir(CO)_2(\eta^5\text{-}C_9H_7)$ to give $WRhIr(CO)_3(\eta\text{-}C_5H_5)(\mu_3\text{-}CTol)(\eta^5\text{-}C_9H_7)_2$. The resulting cluster has a chiral tetrahedral WRhIrC core, and the optical isomers can be resolved by chromatography using a column containing a chiral support.

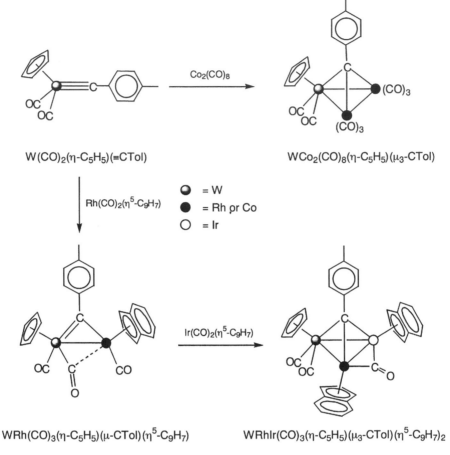

SCHEME 9.4 HETERONUCLEAR CLUSTER FORMATION BY STEPWISE ADDITION ACROSS A M≡C BOND.

### 9.1.4 Redox condensation

Redox condensation involves the reaction between complexes with metals in different formal oxidation states. The term was first applied to the reaction of carbonyl metalates, $[M(CO)_n]^{x-}$, with a neutral metal carbonyl. Redox condensation has been widely used for the synthesis of mixed-metal clusters. A simple example is provided by the reaction of the trinuclear clusters $M_3(CO)_{12}$ (M = Ru, Os) with $[M'(CO)_4]^-$ (M' = Co, Rh, Ir) to generate the tetrahedral clusters $[M_3M'(CO)_{13}]^-$ in good yield:

$$M_3(CO)_{12} + [M'(CO)_4]^- \rightarrow [M_3M'(CO)_{13}]^- + 3CO$$

The square face of the carbide-containing cluster $[Ru_5C(CO)_{14}]^{2-}$ undergoes redox condensation with a number of different metal fragments to give octahedral heteronuclear clusters in excellent yields. An example involves the reaction between $[Ru_5C(CO)_{14}]^{2-}$ with the Group 6 complexes $M(CO)_3L_3$ (M = Cr, L = $NC_5H_5$; M = Mo or W, L = NCMe) in THF to afford the dianionic species $[Ru_5MC(CO)_{17}]^{2-}$ (M = Cr, Mo and W). These clusters have a carbide-centred $Ru_5M$ octahedral core (see Scheme 9.5). The reaction between $[Ru_5C(CO)_{14}]^{2-}$ and $Pt(\eta^4\text{-}C_8H_{12}\text{-}1,5)Cl_2$ affords the heteronuclear cluster $Ru_5PtC(CO)_{14}(\eta^4\text{-}C_8H_{12}\text{-}1,5)$ in 40% yield. Carbonylation of this cluster, by passing CO through a dichloromethane solution of $Ru_5PtC(CO)_{14}(\eta^4\text{-}C_8H_{12}\text{-}1,5)$ at room temperature for a few minutes, affords $Ru_5PtC(CO)_{16}$ in quantitative yield.

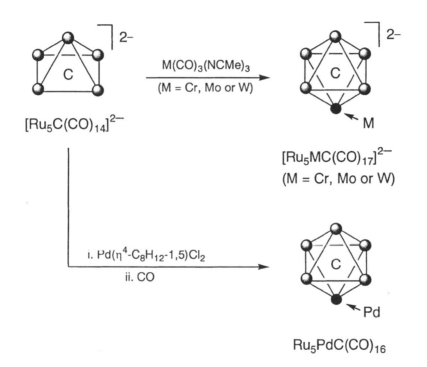

SCHEME 9.5 REDOX CONDENSATION OF MONONUCLEAR FRAGMENTS ACROSS THE SQUARE FACE OF $[Ru_5C(CO)_{14}]^{2-}$.

The reaction of anionic ruthenium clusters with the copper reagent $[Cu(NCMe)_4]Cl$ in $CH_2Cl_2$ leads to the formation of large ruthenium-copper clusters in good yield. The products feature a ruthenium carbonyl cluster on either side of a central unit of copper ions stabilised by bridging chlorine ligands (see Scheme 9.6). The most impressive cluster obtained in this way was isolated from the reaction between $[H_2Ru_{10}(CO)_{25}]^{2-}$ and excess $[Cu(NCMe)_4]Cl$ which affords $[H_4Ru_{20}Cu_6Cl_2(CO)_{48}]^{4-}$. The structure of this cluster can viewed as a polyoctahedral core consisting of six octahrdra, although it is complicated by more extensive metal–metal bonding than this description would imply.

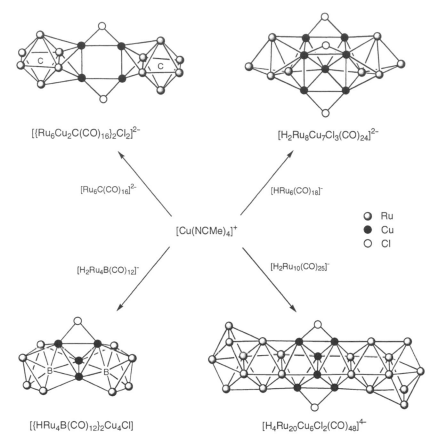

$[\{Ru_6Cu_2C(CO)_{16}\}_2Cl_2]^{2-}$

$[H_2Ru_8Cu_7Cl_3(CO)_{24}]^{2-}$

$[Ru_6C(CO)_{16}]^{2-}$

$[HRu_6(CO)_{18}]^-$

$[Cu(NCMe)_4]^+$

○ Ru
● Cu
○ Cl

$[H_2Ru_4B(CO)_{12}]^-$

$[H_2Ru_{10}(CO)_{25}]^-$

$[\{HRu_4B(CO)_{12}\}_2Cu_4Cl]$

$[H_4Ru_{20}Cu_6Cl_2(CO)_{48}]^{4-}$

SCHEME 9.6  SOME REACTIONS OF $[Cu(NCMe)_4]^+$ WITH RUTHENIUM CLUSTER ANIONS.

A Hydrogen-Storage Model for Palladium metal

The very large bimetallic cluster $H_{12}Pd_{28}(PtPMe_3)(PtPPh_3)_{12}(CO)_{27}$ was isolated from the reduction of $PtCl_2(PMe_3)_2$ and $PdCl_2(PPh_3)_2$ in dimethylsulfoxide at room temperature. The X-ray crystal structure unambiguously determined the positions of all the non-hydrogen atoms, and the metal core is shown below.

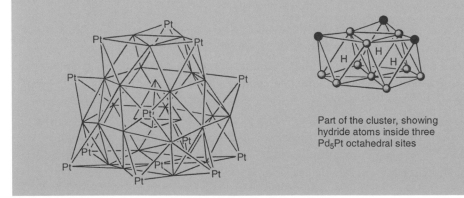

Part of the cluster, showing hydride atoms inside three $Pd_5Pt$ octahedral sites

While no hydrides were located from the X-ray experiment, their presence was suspected due to the large deficiency of electrons based on electron counting methods. Analysis of the $^1H$ NMR spectrum revealed a *pseudo*-triplet resonance at δ −16.4, in a 1:4:1 ratio. This pattern is due to coupling of hydride ligands with one $^{195}Pt$ isotope (spin $^1/_2$, natural abundance 33.8%) to generate a doublet superimposed on a singlet from hydrides displaying no coupling with a nonmagnetic Pt isotope (spin 0, natural abundance 66.2%). The integral of the entire *pseudo*-triplet against the nine protons of the $PMe_3$ ligand showed an 12:9 ratio, providing strong evidence for the presence of 12 hydride ligands and thus satisfying the electron deficiency. The hydrides probably occupy the 12 octahedral-like $Pd_5Pt$ sites.

Porous palladium metal is used for hydrogen storage, purification and isotope separation, and $H_{12}Pd_{28}(PtPMe_3)(PtPPh_3)_{12}(CO)_{27}$ provides a useful model. Similarity in behaviour was demonstrated by treating the cluster with $D_2$ and following the exchange reaction by NMR spectroscopy which showed that 80% of the hydrogens were replaced by deuterium.

### 9.1.5 Bridging ligands

A versatile strategy for the synthesis of mixed-metal clusters is to use ligands capable of forming extra bonds to 'capture' another metal fragment or cluster. While no new metal–metal bonds are formed initially, the proximity of the metal centres means that further ligand elimination usually results in rearrangement to a cluster containing mixed metal–metal bonds.

Bridging chalcogen ligands, with their outwardly pointing lone pair of electrons, are particularly good at capturing metal carbonyl fragments. For example, the sulfido clusters $M_3(\mu_3$-$S)_2(CO)_9$ (M = Fe, Ru, Os) react with the unsaturated fragment $W(CO)_5$, generated by irradiation of $W(CO)_6$ with UV light. Subsequent decarbonylation leads to the formation of W–M bonds, as shown in Scheme 9.7.

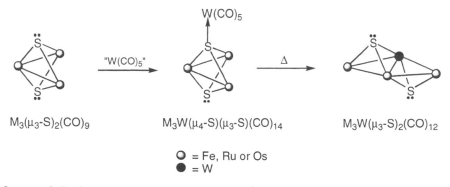

$M_3(\mu_3$-$S)_2(CO)_9$        $M_3W(\mu_4$-$S)(\mu_3$-$S)(CO)_{14}$        $M_3W(\mu_3$-$S)_2(CO)_{12}$

○ = Fe, Ru or Os
● = W

SCHEME 9.7 AN EXAMPLE OF METAL 'CAPTURE' AND SUBSEQUENT MIXED-METAL BOND FORMATION.

The versatility of bridge-assisted syntheses is shown by its ability to combine very different metal fragments to form heteronuclear clusters. For example, the strong π-coordinating ability of an acetylide ligand may be utilised to combine

$W(\eta\text{-}C_5Me_5)(O)_2(CCPh)$, a high oxidation state early transition metal oxo fragment, with $Ru_4(\mu_3\text{-}PPh)(CO)_{13}$, a late transition metal, low oxidation state carbonyl
cluster to afford $Ru_4W(CO)_{10}(\eta\text{-}C_5Me_5)(O)_2(\mu_5\text{-}CCPh)(\mu_4\text{-}PPh)$. The reaction is
shown in Scheme 9.8 and illustrates how the acetylide ligand has become securely
$\pi$-bonded to the $Ru_4$ cluster with the formation of a Ru–W bond.

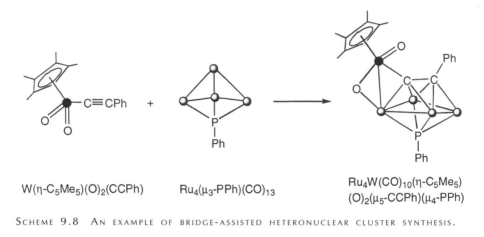

W($\eta$-C$_5$Me$_5$)(O)$_2$(CCPh)    Ru$_4$($\mu_3$-PPh)(CO)$_{13}$    Ru$_4$W(CO)$_{10}$($\eta$-C$_5$Me$_5$)
(O)$_2$($\mu_5$-CCPh)($\mu_4$-PPh)

SCHEME 9.8  AN EXAMPLE OF BRIDGE-ASSISTED HETERONUCLEAR CLUSTER SYNTHESIS.

As well as lone pairs of electrons or $\pi$-donor capability, reactive bonds on ligands
can also be used to connect metal fragments, with subsequent ligand displacement
resulting in the formation of metal–metal bonds. An example is the reaction between
the primary phosphine complex $Cr(CO)_5(PRH_2)$ and $Fe_3(CO)_{12}$, with loss of $H_2$
and CO to form the phosphinidene cluster $CrFe_2(CO)_{11}(\mu_3\text{-}PR)$. The reaction
between $FeCo(CO)_6(\mu\text{-}PRBr)$ and $Na[Mn(CO)_5]$ also results in a phosphinidene
cluster $FeCoMn(CO)_{10}(\mu_3\text{-}PR)$, this time through elimination of NaBr and CO (see
Scheme 9.9).

FeCo(CO)$_6$($\mu$-PRBr)    Na[Mn(CO)$_5$] − NaBr    FeCoMn(CO)$_{10}$($\mu_3$-PR)

SCHEME 9.9  HARNESSING 'REACTIVE BONDS' ON LIGANDS TO INDUCE HETERONUCLEAR
CLUSTER FORMATION.

### 9.1.6 Condensation with electrophilic capping groups

The fragment $[Au(PR_3)]^+$ can be generated from $Au(PR_3)X$ (R = alkyl or aryl; X
= halogen), using thallium(I) salts to extract the halide ion. Analogous copper and
silver fragments are also well known and like $[Au(PR_3)]^+$, are isolobal with $H^+$.
Anionic clusters react with $[Au(PR_3)]^+$ fragments so that the resulting gold unit

bridges cluster edges or face-caps. Addition of two or more equivalents of $[M'(PR_3)]^+$ (M' = Cu, Ag, Au) fragments can lead to the formation of M'–M' bonds as well as M–M' bonds. Some examples illustrating the ways that these fragments bond to clusters are shown in Figure 9.3.

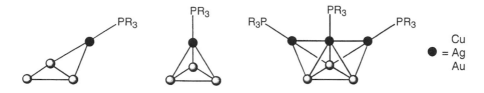

FIGURE 9.3  THE WAYS IN WHICH THE METAL FRAGMENT $[M(PR_3)]^+$ (M=Cu, Ag, Au) MAY BOND TO CLUSTER ANIONS.

### 9.1.7 Serendipitous syntheses

Most syntheses of clusters set out with a product in mind, but in many cases the reaction takes a totally different course. The resulting products are often interesting in their own right, even though the intended synthesis has not gone as anticipated. Scheme 9.10 shows an example of such a reaction.

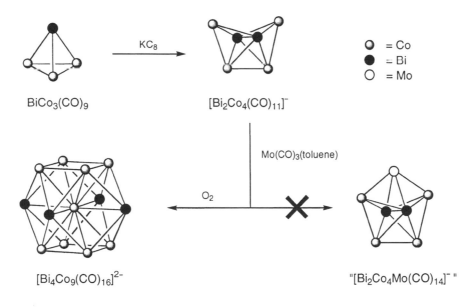

SCHEME 9.10  THE UNEXPECTED SYNTHESIS OF $[Bi_4Co_9(CO)_{16}]^{2-}$.

Addition of a reducing agent, $KC_8$, to the mixed-metal cluster $BiCo_3(CO)_9$, results in the anionic cluster $[Bi_2Co_4(CO)_{11}]^-$. Its structure may be regarded in several different ways, one of which is that of a *nido*-pentagonal bipyramid. The next step in the synthesis was an attempt to form a *closo*-pentagonal bipyramid by adding $Mo(CO)_3$(toluene), which was intended to fill the 'missing' vertex. Instead, no

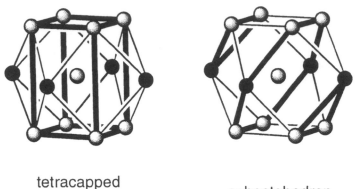

tetracapped
tetragonal prism

cuboctahedron

FIGURE 9.4 THE ALTERNATIVE STRUCTURAL REPRESENTATIONS OF THE METAL CORE IN $[Bi_4Co_9(CO)_{16}]^{2-}$.

Mo–Bi or Mo–Co bonds were formed and the large anionic cluster $[Bi_4Co_9(CO)_{16}]^{2-}$ was generated by oxidation due to the presence of air. The structure of $[Bi_4Co_9(CO)_{16}]^{2-}$, shown in Figure 9.4, can be viewed in two different ways.

If the emphasis is placed on the transition metal atoms, the structure can be regarded as a centred tetragonal prism capped on four faces by the bismuth atoms. However, collectively the metal atoms are close-packed and describe a cuboctahedron. Electron counting formalisms show this cluster to be electron rich, which is explained by a mismatch in the orbitals of Bi versus Co leading to non-bonding orbitals that can accommodate the extra electrons.

## 9.2   HETEROSITE REACTIVITY

Heteronuclear clusters can activate different substrates at different metal atoms. For example, the mixed-metal cluster $Co_2Rh_2(CO)_{12}$ reacts with alkynes at the cobalt atoms, whereas nucleophiles (such as $PR_3$, NCMe) preferentially attach to the rhodium atoms. Further reaction with these substrates causes degradation of the cluster to mixed-metal cobalt-rhodium dimers (see Scheme 9.11).

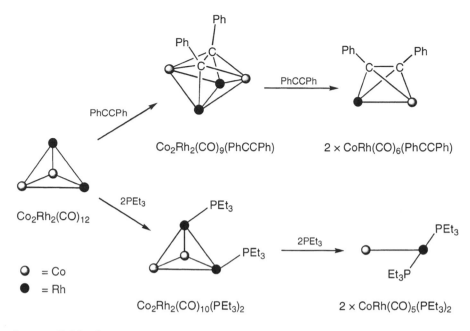

SCHEME 9.11 AN EXAMPLE OF THE DIFFERENCES IN REACTIVITY BETWEEN THE DIFFERENT METALS IN $Co_2Rh_2(CO)_{12}$.

The anionic cluster $[Co_3Ru(CO)_{12}]^-$ reacts with electrophiles to form products in which the $Co_3$ face is capped exclusively (see Scheme 9.12).

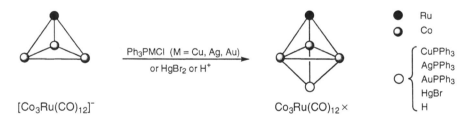

SCHEME 9.12 THE REACTION OF $[Co_3Ru(CO)_{12}]^-$ WITH SOME ELECTROPHILES.

Examples of heteronuclear clusters that catalyse organic reactions are known, and will be described in Chapter 10.

# ORGANIC TRANSFORMATIONS USING STOICHIOMETRIC AND CATALYTIC QUANTITIES OF CLUSTERS

The reactivity of a ligand bonded to a cluster is often different to that of a corresponding ligand bonded to mononuclear complex. There are several reasons for the difference in reactivity; multicentre bonding of the ligand and the ability of the ligand and incoming substrate to approach from different coordination sites being the two most important. This chapter commences with a description of how cluster catalysis can be assessed. It then goes on to give a number of examples and mechanisms for various organic transformations that can be achieved using clusters. Some of these involve stoichiometric reactions although the majority only require catalytic amounts of the cluster to be present. Finally, some methods used to heterogenise clusters by placing them in solid and immiscible liquid supports are described.

## 10.1 ESTABLISHING CATALYSIS BY CLUSTERS

Over the years there has been much debate as to whether cluster catalysts fragment into mononuclear species during reactions or whether they remain intact. It is now well known that examples involving both possibilities exist. In general, however, catalytic reactions involving complete fragmentation of the cluster into mononuclear species are less effective in terms of rate and selectivity than mononuclear catalysts specifically designed for the task. As such, attention will be focused on cluster catalysts that remain intact, although polyhedral transformations involving metal-metal bond breakage and reformation may take place.

It is essential to know what species is actually doing the catalysis as without this knowledge it will not be possible to design better catalysts. Apart from fragmentation of clusters into mononuclear species that are responsible for the catalysis, reaggregation of the fragments into colloids can also occur quite easily and some of these are known to be effective catalysts. It is not an easy task establishing whether a cluster remains intact during catalysis and the first thing to do is find out whether the catalyst is homogeneous or heterogeneous. As well as inspecting the product mixture for solid residue, the following tests should also be carried out:

i. Addition of metallic mercury to a catalyst solution can often indicate if the catalyst is truly homogeneous. The mercury removes colloidal metal particles by forming an alloy and eliminates their contribution to the catalysis. If the activity of the catalyst drops on addition of mercury then catalysis can be assumed to be due to decomposition of the cluster into colloids. However, mercury is known to react with certain clusters so this test alone is not proof and further experiments should be undertaken.

ii. The rigid substrate dibenzocyclooctatetraene coordinates irreversibly to homogeneous catalysts. As such, addition of this substrate to the catalyst solution should prevent catalysis from taking place if the clusters remain in the solution phase.

iii. For catalytic hydrogenation reactions, crosslinked polymers functionalised with alkene groups can be used to assess whether the reaction is catalysed by a homogeneous or heterogeneous catalyst. These polymers can only be hydrogenated by homogeneous catalysts.

Once homogeneous catalysis has been established, the next step is to test whether the catalysis is by the intact cluster or by mononuclear fragments that do not aggregate to form colloids. Sets of tests, based on a large number of detailed experiments, have been developed that help to show if genuine cluster catalysis is taking place. In addition, they are based on more than one metal in the cluster being involved in the catalytic transformation. These tests may be summarised as follows:

i. Increasing the concentration of the catalyst should increase the turnover frequency if the cluster remains intact (turnover frequency is a measure of the moles of substrate converted per unit time per mole of catalyst used).

ii. If the product selectivities obtained using a cluster catalyst are different from those obtained using a mononuclear catalyst then intact cluster catalysis is probably involved.

iii. If heteronuclear (mixed-metal) clusters significantly alter a catalytic process then it is likely that the reaction is catalysed by the intact cluster.

Of course, *in situ* spectroscopic analysis is also very important in establishing the catalytic mechanism. Techniques used to study solutions such as IR and NMR spectroscopy and electrospray ionisation mass spectrometry are all vital in these studies.

## 10.2 HYDROGENATION REACTIONS

Hydrogenation reactions are among the most widely investigated catalysed processes using both homogeneous and heterogeneous catalysts. Mechanistic information is extremely well developed for mononuclear systems and much is also known about the mode of operation of clusters. Large numbers of hydrogenation reactions involving different clusters and substrates are known and some key examples are listed in Table 10.1.

TABLE 10.1   HYDROGENATION OF VARIOUS SUBSTRATES USING CLUSTER CATALYSTS

| Substrate | Product | Catalyst | Conditions | Turnover |
|---|---|---|---|---|
| Cyclohexene | Cyclohexane | $H_4Os_4(CO)_{12}$ | Toluene, 30 atm $H_2$, 150°C, 10 h | 610 |
| Cyclohexene | Cyclohexane | $Pt_2Ir_2(CO)_7(PPh_3)_3$ | Neat, 1 atm $H_2$, 70°C, 5 h | 1600 |
| Styrene | Ethylbenzene | $HCoRu_2(CO)_9(PMe)$ | Cyclohexane, 80 atm $H_2$, 60°C, 45 h | 2720 |
| Hex-1-ene | Hexane (47%) (E)-hex-2-ene (28%) (E)-hex-3-ene (11%) (Z)-hex-2-ene (10%) | $HRu_3(CO)_9(S_2C_4H_8)$ | Toluene, 60 atm $H_2$, 140°C, 1 h | 6910 |
| Benzene | Cyclohexane | $H_2Os_4(CO)_{10}(\eta\text{-}C_6H_6)$ | Hexane, 60 atm $H_2$, 60°C, 6 h | 300 |
| Pent-1-yne | Pent-1-ene (84%) Pent-2-ene (10%) Pentane (3) | $Fe_4(CO)_4(\eta\text{-}C_5H_5)_4$ | Benzene, 7 atm $H_2$, 120°C, 88 h | 340 |

The hydrogenation of alkenes is a very important area in both the chemical and pharmaceutical industries. Mononuclear compounds are extremely effective catalysts for the homogeneous hydrogenation of activated alkenes and clusters have not proved able to compete with them for these types of substrates. Activated alkenes are those containing electron withdrawing functional groups such as alcohols and acids. However, the hydrogenation of non-activated alkenes (simple hydrocarbon alkenes) presents a real problem and clusters are among the most effective homogeneous catalysts available for these substrates.

The hydrogenation of some simple alkenes has been studied using $H_2Os_3(CO)_{10}$. Under certain conditions the hydrogenation of hex-1-ene gives hexane in 30% yield and 70% of internal hexenes. This cluster is unusual in that it only has 46 valence electrons, two fewer than expected for a triangular cluster, and it is believed to possess an Os=Os double bond (see Chapter 3). In the catalytic cycle shown in Scheme 10.1 multiple metal-metal bonds help rationalise the mechanism of the hydrogenation process.

The first step in the catalytic cycle involves association of the alkene to the 46 electron cluster $H_2Os_3(CO)_{10}$, affording the saturated 48 electron cluster $H_2Os_3(CO)_{10}(\eta^2\text{-alkene})$. Association of 2-electron donor ligands to $H_2Os_3(CO)_{10}$ is a well known reaction. With weak ligands such as alkenes the reaction is easily reversed and as such $H_2Os_3(CO)_{10}$ and $H_2Os_3(CO)_{10}(\eta^2\text{-alkene})$ exist in equilibrium. The second step in the cycle involves insertion of a hydride into the metal-alkene

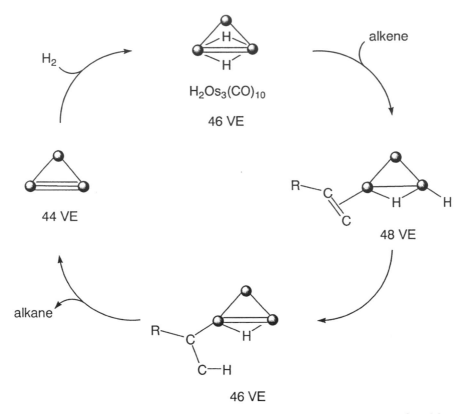

$H_2$

$H_2Os_3(CO)_{10}$

46 VE

alkene

44 VE

48 VE

alkane

46 VE

SCHEME 10.1    THE CATALYTIC CYCLE FOR THE HYDROGENATION OF ALKENES USING $H_2Os_3(CO)_{10}$.

bond to give the 46 valence electron (VE) alkyl derivative $HOs_3(CO)_{10}(\eta^1\text{-alkyl})$. The fact that hydrogenation using $H_2Os_3(CO)_{10}$ results in the formation of internal hexenes in high yield compared with saturated hydrogenation products indicates that the insertion and association steps are reversible. The most speculative part of this cycle is the formation of the 44 VE intermediate $Os_3(CO)_{10}$ that is formed on dissociation of the alkane. This intermediate is believed to have an Os≡Os triple bond and would readily react with molecular hydrogen to regenerate the starting cluster $H_2Os_3(CO)_{10}$.

The catalytic hydrogenation cycle described above differs from that proposed for most clusters as it commences with a coordinatively unsaturated cluster. The majority of clusters are saturated and as such the association step must be accompanied by either dissociation of a suitable ligand or by metal-metal bond breaking. An example of a more commonly encountered mechanism involving ligand dissociation as the initial step in the catalysis cycle is provided by $H_4Ru_4(CO)_{12}$ (see Scheme 10.2). Examples of catalytic processes invoking metal-metal bond scission (i.e. polyhedral rearrangements) will be given in Section 10.3.

The first step in the catalytic cycle for the $H_4Ru_4(CO)_{12}$ catalysed hydrogenation of alkenes is the dissociation of a CO ligand. This is required because $H_4Ru_4(CO)_{12}$ is a coordinatively saturated 60 electron cluster with electron precise (i.e. two centre/

two electron) Ru-Ru bonds. The $H_4Ru_4(CO)_{11}$ intermediate has 58 valence electrons and a vacant coordination site, which may accept the incoming alkene ligand to afford the 60 electron species $H_4Ru_4(CO)_{11}(\eta^2\text{-alkene})$. A competing reaction is the oxidative addition of $H_2$ since the reaction takes place under a $H_2$ atmosphere.

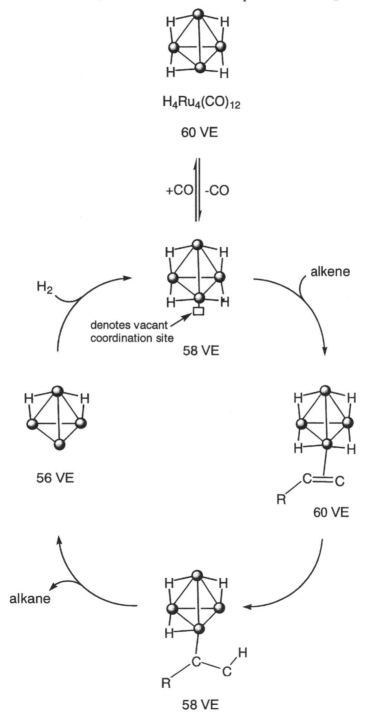

SCHEME 10.2  THE CATALYTIC CYCLE FOR THE HYDROGENATION OF ALKENES USING $H_4Ru_4(CO)_{12}$.

In the next step (similar to the previous example), insertion of a metal hydride into the metal-alkene bond occurs giving the 58 electron alkyl-coordinated intermediate $H_3Ru_4(CO)_{11}(\eta^1\text{-alkyl})$. The next step shown in the Scheme involves loss of the alkane to give a 56 VE intermediate that rapidly takes up $H_2$ to form the 58 electron cluster with the vacant coordination site. The intermediates proposed fit the experimental observations well. Dissociation of CO is the rate-determining step in the reaction sequence as it has the highest activation barrier. Replacing one or more of the CO ligands with phosphines can lower the activation barrier. The phosphine forms a weaker bond to the cluster, dissociates more readily and lowers the conditions required to conduct the catalytic process. It is also possible that the alkene bonds to the cluster in other multicentre bonding modes than the $\eta^2$-coordination mode shown in the Scheme. However, when catalyst activation is by CO loss (as opposed to metal-metal bond cleavage), only one metal atom usually participates in bonding to the substrate.

## Catalytic reduction of arenes

Removal of benzene and other aromatic compounds from diesel fuel is an important environmental concern due to the carcinogenic nature of many of these compounds. One of the best ways to remove such compounds is not to extract them from the other fuel components by expensive fractional distillation methods, but to hydrogenate them to saturated ring compounds that are comparatively harmless. This process is achieved in industry using heterogeneous catalysts although an effective homogeneous catalyst would be desirable for certain processes. There are two main difficulties to overcome. First, aromatic rings are notoriously difficult to hydrogenate due to the extra stabilisation energy the delocalised bonding provides, and second, fuels tend to contain heteroaromatic impurities that poison mononuclear catalysts rapidly preventing them from working. Clusters provide an ideal solution to this latter problem and two homogeneous cluster catalysts have been found to effectively hydrogenate benzene to cyclohexane.

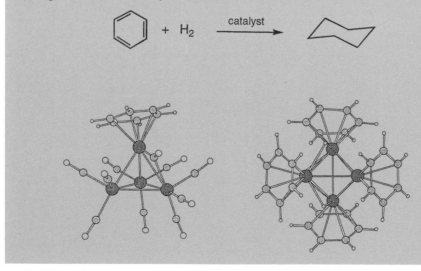

The tetraruthenium cluster $[H_4Ru_4(\eta\text{-}C_6H_6)_4]^{2+}$ has only 58 valence electrons and reacts with $H_2$ to give $[H_6Ru_4(\eta\text{-}C_6H_6)_4]^{2+}$, the active catalyst species. This cluster is soluble in water and ionic liquids and is used in biphasic processes (see Section 10.7.2). The tetraosmium cluster $H_2Os_4(CO)_{10}(\eta\text{-}C_6H_6)$ is soluble in polar organic solvents typically encountered in homogeneous catalytic reactions. Both clusters are presumed to operate by the same mechanism, involving stepwise slippage of the ring as each unsaturated bond of the aromatic ring is hydrogenated.

## 10.3 HYDROFORMYLATION

Hydroformylation is the name given to the reaction involving addition of hydrogen and carbon monoxide to an unsaturated organic substrate. Hydroformylation of alkenes yields mostly aldehydes and the main challenge in this area is to obtain one product, either the *n*- or *iso*-aldehyde (see Scheme 10.3) in the highest yield possible. In addition, the reaction can go further such that the aldehyde converts to an alcohol. Hydroformylation catalysts must be regioselective, producing either the *n*- or *iso*-compound, and chemically selective, producing either an aldehyde or an alcohol. Mixtures of products means that expensive separation procedures are needed and certain products may not be wanted, which clearly leads to waste of materials and energy.

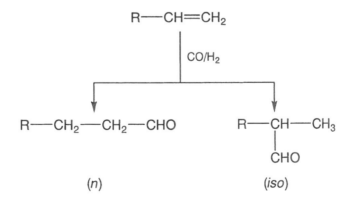

SCHEME 10.3

One cluster that has been found to be a particularly good hydroformylation catalyst is $Rh_4(CO)_{12}$. It has been used for the hydroformylation of many alkene substrates and is often used in combination with a range of phosphine ligands that help to modify its selectivity. A catalytic cycle has been proposed for the hydroformylation of alkenes using $Rh_4(CO)_{12}$ that involves metal-metal bond breaking and formation. This cycle is illustrated in Scheme 10.4 and is based on reactions that are known to occur under moderate pressures of CO and $H_2$. Under high pressure there is evidence to suggest that complete fragmentation of the $Rh_4(CO)_{12}$ cluster takes place.

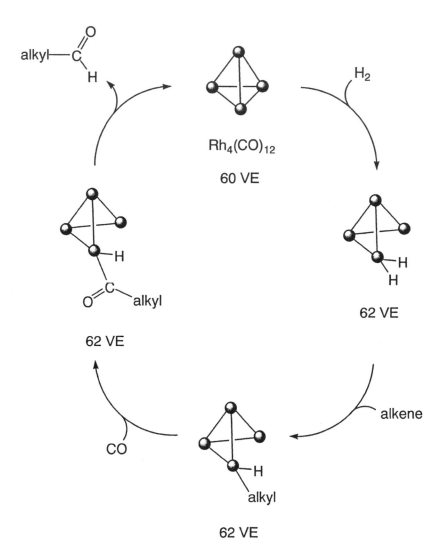

$Rh_4(CO)_{12}$

60 VE

62 VE

62 VE

62 VE

SCHEME 10.4 THE CATALYTIC CYCLE FOR THE HYDROFORMYLATION OF ALKENES USING $Rh_4(CO)_{12}$.

The first step in the catalytic cycle involves oxidative addition of hydrogen to $Rh_4(CO)_{12}$ with simultaneous breaking of a Rh-Rh bond. The valence electron count of the hydride-cluster is 62 as expected for a butterfly. The process is easily reversible, but in the presence of an alkene substrate, a Rh-alkyl intermediate is produced. There is no evidence for the formation of an intermediate involving a coordinated alkene, although this is quite likely. It would require either a second Rh-Rh bond to break or dissociation of a carbonyl ligand. Since the reaction takes place under a pressure of CO both steps would be extremely rapid and it is therefore not surprising that such an intermediate cannot be detected. The next step in the cycle involves insertion of a carbonyl group into the Rh-alkyl bond and could also involve two steps. First, insertion of a coordinated CO that would result in closure of the cluster to a tetrahedron, and second, uptake of CO to regenerate a 62 electron butterfly intermediate. In the final step, the aldehyde is eliminated from the cluster to regenerate $Rh_4(CO)_{12}$.

The conversion of diphenylacetylene into α-phenylcinnamaldehyde using a cluster hydroformylation catalyst is shown in Scheme 10.5. This reaction has been studied in some detail and the two species shown in the scheme have been actually isolated, other intermediates that are not shown can also be postulated. The starting compound is $HRu_3(CO)_9(\mu_3\text{-}\eta^2\text{-NMePy})$ and the face-capping 2-(methylamino)pyridyl ligand helps confer stability to the cluster core preventing it from breaking down during reaction.

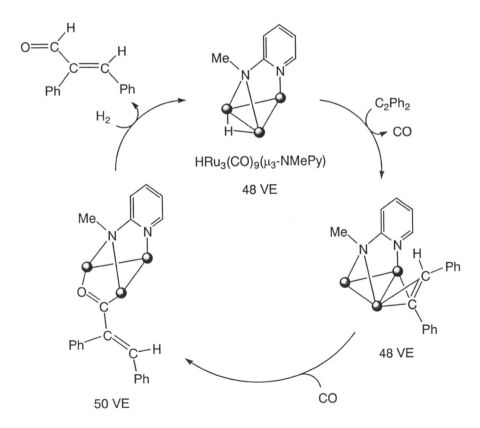

SCHEME 10.5 THE CATALYTIC CYCLE FOR THE CONVERSION OF DIPHENYLACETYLENE INTO α-PHENYLCINNAMALDEHYDE USING $HRu_3(CO)_9(\mu_3\text{-}\eta^2\text{-NMePy})$.

The first step in the catalytic cycle involves loss of CO and uptake of diphenylacetylene, which adopts a $\mu$-$\eta^2$-bonding mode as shown in the Scheme. It is likely that this step initially involves the breaking of a Ru-Ru bond, association of the alkyne, followed by elimination of a CO ligand and then reformation of the Ru-Ru bond. In the next step migratory insertion of the alkyne takes place producing a 50-electron cluster in which one Ru-Ru bond has broken. The inserted CO straddles this open Ru-Ru edge and donates 4-electrons. This cluster undergoes attack by $H_2$ with hydrogenolysis of both the cluster and propenoyl group to give $\alpha$-phenylcinnamaldehyde and the 48 electron precursor cluster $HRu_3(CO)_9(\mu_3$-$\eta^2$-NMePy) after reformation of the Ru-Ru bond.

The process described above is not without problems and depending upon the conditions employed the active catalyst can be poisoned. Conversion of $HRu_3(CO)_9(\mu_3$-$\eta^2$-NMePy) to the catalytically inactive dinuclear species $Ru_2(CO)_6(\eta^3$-NMePyCO) takes place after several cycles. The dimer is formed by insertion of CO into the Ru-amide bond.

## 10.4 THE WATER-GAS SHIFT REACTION

Hydrogenation and hydroformylation reactions can be carried out using water as the source of $H_2$. The conversion of carbon monoxide and water into carbon dioxide and $H_2$ is known as the water-gas shift reaction. It can be carried out on its own using cluster catalysts to generate $H_2$ or in the presence of alkenes to generate alkanes or aldehydes.

Some of the most effective homogeneous catalysts for the water-gas shift reaction are anionic ruthenium clusters. The main active catalytic species generated from $Ru_3(CO)_{12}$ in alkaline aqueous-alcoholic solutions under an atmospheric pressure of CO are $[HRu_3(CO)_{11}]^-$ and $[H_3Ru_4(CO)_{12}]^-$. Based on a considerable amount of experimental evidence, the mechanistic cycle shown in Scheme 10.6 has been devised to account for the conversion of CO and $H_2O$ into $CO_2$ and $H_2$. The cycle commences with the activation of $Ru_3(CO)_{12}$ by $OH^-$ to afford $[HRu_3(CO)_{11}]^-$ and $CO_2$. Catalytic amounts of $OH^-$ are required to initiate the reaction, and $OH^-$ is regenerated at a later stage in the reaction cycle. The first step already accounts for the formation of the $CO_2$ in the process, but the $H_2$ still needs to be produced. In the second step of the cycle association of CO to the 48 electron anionic intermediate results in breaking one of the Ru-Ru edges to afford the 50 electron cluster $[HRu_3(CO)_{12}]^-$. This cluster readily reacts with $H^+$ obtained from the scission of water to give the neutral 50 electron species $H_2Ru_3(CO)_{12}$. This species would be very short lived and can revert back to the starting compound $Ru_3(CO)_{12}$ by reductive elimination of $H_2$.

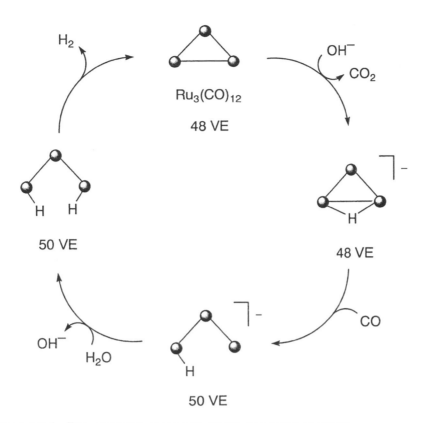

SCHEME 10.6 THE CATALYTIC CYCLE FOR WATER-GAS SHIFT REACTION.

The ability to generate $H_2$ from water is an extremely useful process in itself, however, being able to make new organic compounds *in situ* is also highly desirable. An example of a hydrogenation reaction where this can be achieved is for the selective hydrogenation of quinoline (see Scheme 10.7). The nitrogen containing ring is hydrogenated using carbon monoxide and water in the presence of catalytic amounts of $Rh_6(CO)_{16}$.

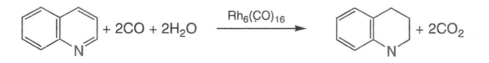

SCHEME 10.7 THE SELECTIVE HYDROGENATION OF QUINOLENE UNDER THE WATER-GAS SHIFT REACTION CONDITIONS.

An example of hydroformylation using water as the source of $H_2$ employed on an industrial scale is the conversion of propene to butanol. The process is known as the Reppe synthesis. The reaction uses iron carbonyls as the catalysts and while species such as $[HFe_3(CO)_{11}]^-$ have been isolated from the reaction, the active catalysts is believed to be the mononuclear species $[HFe(CO)_4]^-$.

## 10.5  C-H BOND ACTIVATION AND C-C BOND FORMATION

There are vast numbers of cluster reactions known that involve C-H bond activation and quite a number of these are catalytic. Only a couple of examples involving regioselective acylation of heterocyclic rings will be given here. The examples have been chosen because of their usefulness in organic synthesis. The reactions in question afford compounds that cannot be made by other routes without considerable difficulty and the superiority of the cluster catalyst is because activation of the organic substrate by the cluster involves a multicentre interaction. Both the acylation reactions are catalysed by $Ru_3(CO)_{12}$ and are illustrated in Scheme 10.8.

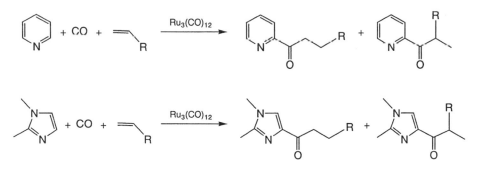

SCHEME 10.8   THE SELECTIVE ACYLATION OF HETEROCYCLES CATALYSED BY $Ru_3(CO)_{12}$.

The first reaction shown in Scheme 10.8 involves C-H bond activation at the *ortho*-carbon atom in pyridine. This regioselectivity is possible as the nitrogen atom bonds to one metal centre and the carbon atom to a different metal centre. The structure of the proposed intermediate is shown in Figure 10.1.

The second example in Scheme 10.8 involves the acylation of an imidazole ring and a related intermediate to that shown in Figure 10.1 can be envisaged. Activation takes place at the *ortho*-carbon atom to the nitrogen atom that coordinates to the cluster. The reaction works for a wide range of substrates including alkyl, aryl and silyl groups as well as a number of other heterocycles and two further examples are given in Scheme 10.9. The site of the C-H bond activation can be appreciated relative to the location of the nitrogen atom donor. The activity of $Ru_3(CO)_{12}$ in these reactions is unique. A number of other clusters such as $Fe_3(CO)_{12}$, $Os_3(CO)_{12}$ and $Rh_4(CO)_{12}$ have been tested and found not to catalyse these acylation reactions.

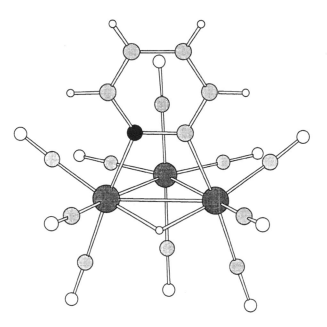

FIGURE 10.1 THE STRUCTURE OF $HRu_3(CO)_{10}(NC_5H_4)$.

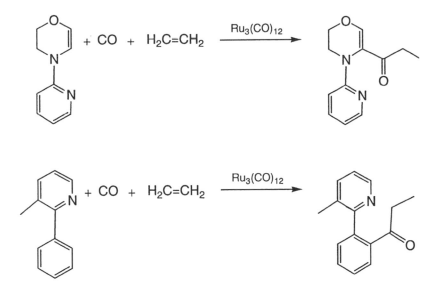

SCHEME 10.9 THE SELECTIVE ACYLATION OF HETEROCYCLES CATALYSED BY $Ru_3(CO)_{12}$.

## Dehydrosulfurisation by $Ru_5C(CO)_{15}$

Removal of sulfur from fuels is important for environmental reasons. When sulfur compounds are burnt $SO_3$ is produced, which reacts with water causing acid rain. Removal of the sulfur takes place in two steps. First, the sulfur is extracted from the organosulfur compounds as hydrogen sulfide ($H_2S$), and second, dehydrosulfurisation of $H_2S$ to elemental sulfur must be undertaken, as $H_2S$ is toxic. This latter reaction is conducted on a large scale using the Claus process according to the equation below:

$$2\ H_2S + SO_2 \rightarrow 3/8\ S_8 + 2\ H_2O$$

The Claus process achieves this very efficiently using a heterogeneous alumina catalyst, which operates at temperatures in excess of 300°C. An alternative method has been developed that generates sulfur from $H_2S$ under ambient conditions using a pentaruthenium cluster.

The square pyramidal cluster $Ru_5C(CO)_{15}$ reacts rapidly with $H_2S$ at room temperature to afford $HRu_5C(CO)_{14}(\mu\text{-SH})$ as shown in the Scheme. The metal polyhedron of $HRu_5C(CO)_{14}(\mu\text{-SH})$ is based on a bridged butterfly (as expected for a cluster with an electron count of 76) and the $\mu$-SH unit bridges a hinge ruthenium atom and the bridging Ru-atom. The hydride ligand bridges across the Ru-Ru hinge bond.

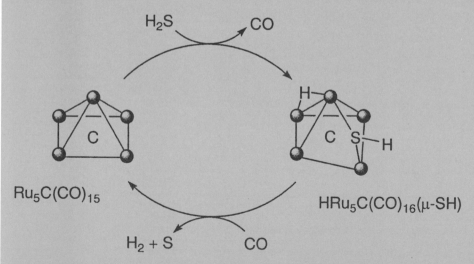

$Ru_5C(CO)_{15}$           $HRu_5C(CO)_{16}(\mu\text{-SH})$

Reaction of $HRu_5C(CO)_{14}(\mu\text{-SH})$ with CO at atmospheric pressure results in the regeneration of $Ru_5C(CO)_{15}$ and the formation of elemental sulfur and $H_2$ gas. While this process is not catalytic it represents a simple homogeneous method for dehydrosulfurisation under ambient conditions based on a facile cluster structural rearrangement. No CO is consumed in the process as one equivalent is lost in the first step, and one equivalent is reintroduced in the second step.

Clusters are widely used to catalyse nucleophilic addition to alkynes (see Scheme 10.10). As for the other C-H bond activation reactions, the cluster of choice is $Ru_3(CO)_{12}$ and the three reactions illustrated are:

i.   Regioselective nucleophilic addition with carboxylic acids to terminal alkynes.
ii.  Attack of an alkyne at the $C_2$ position by formic acid ($HCO_2H$) to afford an enol ester with an *exo*-alkene.
iii. The reaction between dimethylalkynedicarboxylate and formic acid to afford an $\alpha$-pyrone ring.

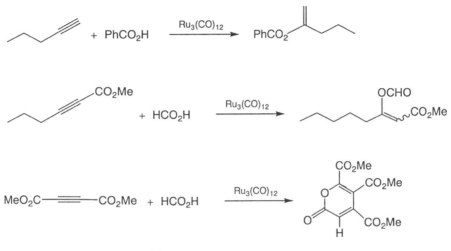

SCHEME 10.10 SOME $Ru_3(CO)_{12}$ CATALYSED NUCLEOPHILIC ADDITION REACTIONS.

The mechanism for these reactions has not been established, but since similar transformations can be achieved by a number of mononuclear catalysts, it is unlikely that multicentre bond activation is involved.

## 10.6 MIXED-METAL CLUSTERS IN CATALYSIS

A large number of mixed-metal clusters have been used in catalysis but comparisons with related homonuclear counterparts are often not made. However, there are some examples where it has been demonstrated that heteronuclear clusters catalyse reactions differently and more efficiently than related homonuclear clusters. For example, the three clusters $Rh_4(CO)_{12}$, $Rh_2Co_2(CO)_{12}$ and $RhCo_3(CO)_{12}$ catalyse the hydrosilylation of 1-hexyne with $HSiEt_3$ (see Scheme 10.11). Under identical conditions, all the reactions proceed with 100% conversion of the substrate. The homonuclear cluster $Rh_4(CO)_{12}$ is the least regioselective of the three catalysts while $RhCo_3(CO)_{12}$ is the most regioselective. One might expect that as the ratio of cobalt increases in the catalyst, then $Co_4(CO)_{12}$ would give the highest regioselectivity. However, $Co_4(CO)_{12}$ is by far the least selective and decomposes during reaction.

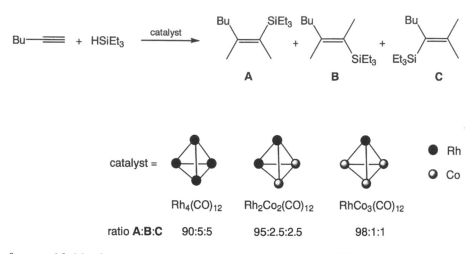

SCHEME 10.11   A COMPARISON OF REGIOSELECTIVITY OF $M_4(CO)_{12}$ CATALYSTS.

There are many other examples like the one shown above where a mixed-metal cluster is more regioselective than a homonuclear cluster catalyst. The mechanism for these reactions is not well understood and the reason for the different selectivities is often difficult to rationalise. Experiments on $Rh_4(CO)_{12}$ show that two different catalytically active species are present, possibly due to fragmentation, and the two species are responsible for giving either the $Z$ product (A) or the $E$ products (B and C). With the heteronuclear clusters, an even greater number of different active catalytic species could be formed during reaction, by virtue of containing two different types of metals, and this factor could account for their differences in regioselectivity.

A high nuclearity heteronuclear cluster that can catalyse various alkyne transformations without undergoing fragmentation is $H_4Pt_3Ru_6(CO)_{21}$. The structure of this cluster is based on three trimetal layers in which a $Pt_3$ triangle is sandwiched between two $Ru_3$ layers (see Scheme 10.12). Each end of the cluster is bridged by hydride ligands; three edge bridging at one end and one face-capping the other. This cluster reacts with $PhC \equiv CPh$ to afford $H_2Pt_3Ru_6(CO)_{20}(\mu_3\text{-}C_2Ph_2)$, which is the active catalyst for the hydrogenation and hydroslylation of alkynes. The cluster is superior to its homonuclear analogues for the hydrogenation of diphenylacetylene to $(Z)$-stilbene although catalyst recovery is not quantitative and the catalyst lifetime is limited.

The catalytic transformations in $H_2Pt_3Ru_6(CO)_{20}(\mu_3\text{-}C_2Ph_2)$ occur at the $Ru_3$ faces and the platinum metals do not appear to be actively involved. The first step in the catalytic cycle involves dissociation of a CO ligand together with oxidative addition of $H_2$. The CO is believed to be lost from a ruthenium atom, and while the location of the new hydrides is unknown, on the scheme both have been shown bonding to the same ruthenium atom. Association of a second alkyne is thought to take place next, accompanied by insertion of a hydride into a Ru-C bond of the original alkyne. As the bonding mode of the new alkyne changes from a simple two-

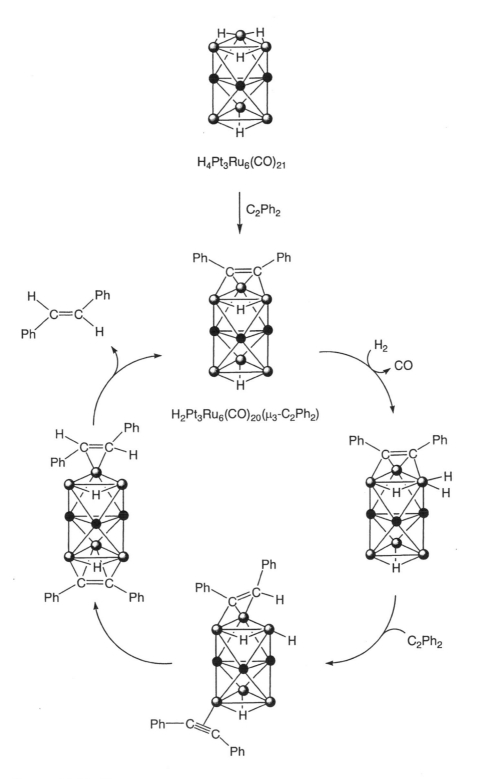

SCHEME 10.12 THE CATALYTIC CYCLE FOR THE CONVERSION OF DIPHENYLACETYLENE TO (Z)-STILBENE USING A HETERONUCLEAR LAYERED CLUSTER CATALYST.

electron donor interaction to a four electron face-capping interaction, the second insertion of a hydride takes place to give the $\eta^2$-coordinated alkene. Loss of the alkene results in the regeneration of the starting cluster, with the new alkyne substrate on the opposite $Ru_3$ face to the original one.

Apart from catalysing the hydrogenation of diphenylacetylene to (Z)-stilbene, $H_2Pt_3Ru_6(CO)_{20}(\mu_3-C_2Ph_2)$ also catalyses the hydrosilation of diphenylacetylene. The catalytic cycle differs in that the $HSiEt_3$ adds to the cluster after dissociation of a CO has occurred and the second alkyne coordinates to the same face as the original one. It is possible that this is also the case in the hydrogenation reaction. However, different mechanisms are probably in operation as diphenylacetylene is converted to (E)-$Ph(H)C=C(SiEt_3)Ph$ rather than the Z-isomer. The question that must be addressed is whether the enhanced catalytic activity of the bimetallic cluster, compared with homonuclear clusters, is due to the presence of the platinum layer as it does not appear to be directly involved in the reaction. However, the platinum layer does have a pronounced effect transmitted to the ruthenium layers to give what has been described as a 'synergistic enhancement' in which the platinum layer donates electron density to the ruthenium layers.

## 10.7 SUPPORTED CLUSTER CATALYSIS AND USE OF NON-ORGANIC SOLVENTS

Homogeneous catalysts are desirable as they often operate under mild conditions and product selectivity is high. The main disadvantage of homogeneous catalysis is the problem of recovering the expensive catalyst once the reaction is complete. A heterogeneous catalyst can be anchored to a reaction vessel or may be removed simply by filtration. In contrast, a homogeneous catalyst is in the same phase as the product and removal is time consuming and costly. With this in mind a number of approaches have been devised, which effectively heterogenise a homogeneous catalyst so that they become easier to handle and make separation after reaction relatively straightforward.

Another motivation for supporting metal clusters on solids is that many industrial catalysts consist of metal nanoparticles dispersed on porous solid supports. The smaller the nanoparticle the larger the fraction of metal atoms that are exposed at the surface and therefore able to participate in catalysis. One way of preparing supports containing tiny groups of metal atoms is to deposit a cluster on the support and then convert it to a metal particle, usually by heating at a high temperature to drive off all the ligands. These naked metal particles are sometimes preferred to actual tethered molecular clusters as they are more robust and strongly bound to the support and can therefore be reused indefinitely.

### 10.7.1 Solid supported clusters

The most common materials used to support metal clusters are metal oxides, zeolites and carbon. While the goal is to produce a naked metal particle, the best way to achieve this is to deposit a molecular compound in the first instance. Ideally, the compound should be of known nuclearity and when converted to the naked metal particle, a uniform particle size and distribution throughout the porous support should be achieved. However, straightforward deposition of clusters is relatively unsuccessful, as they do not deposit in a uniform manner, which means that aggregation takes place during conversion to naked metal particles. A way to circumvent this problem is to commence with a mononuclear complex and prepare the cluster on the surface. A series of reactions are shown in Scheme 10.13 that illustrate the stepwise synthesis of some iridium clusters on alumina ($\gamma$-Al$_2$O$_3$). The exact product is dependent on the basicity of the $\gamma$-Al$_2$O$_3$. These reactions can also be conducted on magnesium oxide surfaces as well as in zeolites. The preparation of high nuclearity clusters from mononuclear precursors inside a zeolite cage is particularly attractive if the cluster produced is larger than the pores in the zeolite. Substrates may enter the zeolite and the products may be extracted but the cluster remains inside. These systems are often referred to as 'ship in the bottle' or 'teabag' catalysts.

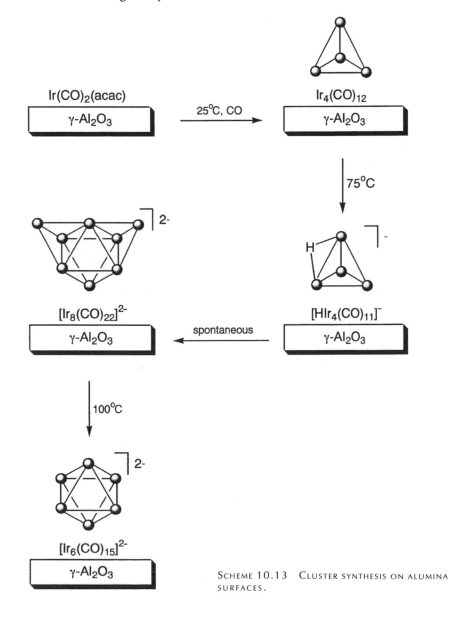

SCHEME 10.13   CLUSTER SYNTHESIS ON ALUMINA SURFACES.

The clusters shown in Scheme 10.13 can be removed from the alumina by extraction into polar organic solvents containing appropriate counter ions. Alternatively, decarbonylation of the clusters by heating them to very high temperatures under an inert atmosphere affords small, well-defined naked iridium particles. Structural rearrangements can often accompany the decarbonylation process and methods such as decarbonylation in the presence of $H_2$ reduce the temperature at which the conversion takes place and helps to prevent these changes from occurring. The iridium cluster-alumina catalysts have been used to catalyse a number of different reactions, notably, the hydrogenation of alkenes and arenes.

### 10.7.2 Biphasic 'liquid-liquid' supports

Biphasic catalysis is an alternative method used to heterogenise a homogeneous catalyst. Instead of supporting the catalyst on a solid where there is a possibility that its activity or selectivity may be affected, the catalyst is supported in a liquid phase that is immiscible with the substrate/product phase. In aqueous-organic biphasic catalysis, for example, the catalyst is supported (dissolved) in water and the substrate dissolved in an immiscible organic solvent (sometimes neat substrate is used). The two phases are then vigorously agitated so that they mix and the catalyst brings about conversion to the product. Once reaction is complete, agitation is stopped and the two phases separate. This process is shown schematically in Figure 10.2. In principle, the product phase can simply be decanted or drained from the aqueous phase containing the catalyst.

Most clusters are soluble in organic solvents and are not suited to an aqueous environment. In order to induce water solubility and inhibit solubility in organic solvents appropriate ligands must be attached to the cluster. A wide range of water soluble phosphine ligands have been developed for this purpose and three examples are given in Figure 10.3.

One ligand that has been used to render clusters water soluble is *tris*-(3-sulfonatophenyl)phosphine trisodium salt, TPPTN, (top left in the figure). Preparation of clusters with this ligand can be achieved in two ways:

i. Direct reaction in a solvent such as methanol.

ii. Biphasic ligand exchange reactions where a cluster with, say triphenylphosphine, is dissolved in an organic solvent and stirred vigorously with an aqueous solution of the water soluble ligand. As the triphenylphosphine is exchanged with the water soluble ligand the cluster is drawn into the aqueous phase.

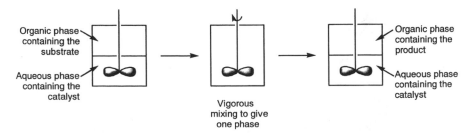

Organic phase containing the substrate

Aqueous phase containing the catalyst

Vigorous mixing to give one phase

Organic phase containing the product

Aqueous phase containing the catalyst

FIGURE 10.2 REPRESENTATION OF THE BIPHASIC APPROACH.

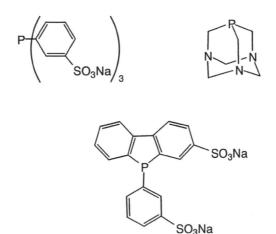

FIGURE 10.3 SOME WATER SOLUBLE PHOSPHINE LIGANDS.

Clusters such as $Ru_3(CO)_{12-x}(TPPTN)_x$ (x = 1 - 3) and $H_4Ru_4(CO)_{12-x}(TPPTN)$ (x = 1 − 4) have been prepared using these methods. The related non-aqueous ruthenium clusters have been shown to act as hydrogenation and hydroformylation catalyst (or catalyst precursors) and the related water-soluble species catalyse similar reactions under biphasic liquid-liquid conditions.

There are other biphasic systems available apart from aqueous-organic biphasic catalysis, such as fluorous phase-organic and ionic liquid-organic. In addition, catalysis in other non-organic solvents are being developed as a clean way to conduct synthesis. An example of a non-organic solvent that is widely used in catalysis is supercritical carbon dioxide ($scCO_2$). Supercritical fluids, in general, are attractive solvents as they form single-phase mixtures with gaseous reactants, which can lead to enhanced reaction rates. Supercritical $CO_2$ is a particularly desirable solvent as it is abundant, inexpensive, nonflammable, non-toxic and environmentally benign. In addition, nonpolar organic compounds are highly soluble in $scCO_2$. Catalysis by clusters in supercritical fluids is poorly developed. However, $Fe_3(CO)_{12}$ has been used in $scCO_2$ to catalyse the isomerisation of 1-hexene to 2-hexene as shown in Scheme 10.14. The reaction takes place in high yield, and rapidly, unfortunately higher temperatures are required than the corresponding reaction in the neat alkene. The reason for this remains unclear, however, further study of transition metal carbonyl clusters in $scCO_2$ will continue as they are considerably more soluble than many other catalysts currently available.

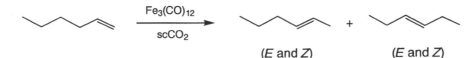

SCHEME 10.14 $Fe_3(CO)_{12}$ CATALYSED ISOMERISATION OF 1-HEXENE TO 2-HEXENE IN SUPERCRITICAL $CO_2$.

# Further Reading

A number of books about transition metal clusters have been written, the vast majority aimed specifically at researchers working in the area. Some key books and review articles that may be of interest to the reader who wishes to delve deeper into transition metal carbonyl cluster chemistry are listed below.

## GENERAL TEXTS

'An introduction to cluster chemistry', D. M. P. Mingos and D. J. Wales, Prentice-Hall, London, 1990.

'The synergy between dynamics and reactivity at clusters and surfaces', Ed. L. J. Farrugia, Kluwer, Dordrecht, 1994.

'Metal-metal bonded carbonyl dimers and clusters', C. E. Housecroft, Oxford University Press, Oxford, 1996.

An entire issue of *Coord. Chem. Rev.* was devoted to cluster chemistry, 1995, Volume 143, pages 1–678.

An entire issue of *J. Chem. Soc., Dalton Trans.* was devoted to cluster chemistry, 1996, Number 5, pages 555–800.

## ELECTRON COUNTING AND STRUCTURE

'Structural systematics in nickel carbonyl cluster anions', A. F. Masters and J. T. Meyer, *Polyhedron*, 1995, **14**, 339.

'Electron counting in clusters: a view of the concepts', S. M. Owen, *Polyhedron*, 1998, 7, 253.

## SYNTHESIS AND REACTIVITY

'High-nuclearity carbonyl clusters: their synthesis and reactivity', M. D. Vargas and J. N. Nicholls, *Advs. Inorg. & Radiochem.*, 1986, **30**, 123.

## HETERONUCLEAR CLUSTERS

'Main group-transition metal cluster compounds of the Group 15 elements', K. H. Whitmire, *Advs. Organomet. Chem.*, 1998, **42**, 1.

## CATALYSIS

'Transition metal clusters in homogeneous catalysis', G. Süss-Fink and G. Meister, *Advs. Organomet. Chem.*, 1993, **35**, 41.

'Catalysi by di- and polynuclear metal cluster complexes', Eds. R. D. Adams and F. A. Cotton, Wiley-VCH, New York, 1998.

# Index

# Cluster Formula Index